Principles of Valuation

John Armatys, Phil Askham
and Mike Green

2009

A division of Reed Business Information

Estates Gazette
1 Procter Street, London WC1V 6EU

EG Books is an imprint of Elsevier
The Boulevard, Langford Lane, Kidlington, Oxford OX5 1GB, UK
30 Corporate Drive, Suite 400, Burlington, MA 01803, USA

First edition 2009
Reprinted 2010

British Library Cataloguing in Publication Data
A catalogue record for this book is available from the British Library

Library of Congress Cataloging-in-Publication Data
A catalog record for this book is available from the Library of Congress

ISBN: 978-0-7282-0568-0

For information on all EG Books publications
visit our website at www.elsevierdirect.com

Printed and bound in *Great Britain*

10 11 12 13 14 15 10 9 8 7 6 5 4 3 2

Contents

Acknowledgements

Writing a text such as this is a significant undertaking which would not be possible without the efforts of a large number of people, too many to mention perhaps but the authors would like to thank, in particular:

- Sue Mason for her drafting skills in providing the illustrations for the book
- Cushman and Wakefield LLP for supplying data for Figure 7.1 on the commercial property market
- John Strawinski and Kevin Howes for their assistance with Chapter 10 on the profits method
- EG Books Ltd's editorial and production staff, in particular Alison Bird and Sarah Jackman, for their valued assistance in the production of this book
- our families for their encouragement and assistance
- those who taught us when we were students
- colleagues and friends in the property and education professions for the knowledge, information and ideas included in this book
- and finally thanks are due for the indirect contribution of the many hundreds of students who have passed through our courses and helped us to develop our understanding of how best to explain and illustrate some of the more complex aspects of valuation methodology.

Introduction

This book provides a readable and practical guide to property valuation methodology. It is aimed at first year undergraduate students on property courses and students on post graduate conversion courses recognised by the RICS. We have started from first principles and it is our intention that the book should stand alone as a self-contained guide to the subject. It is assumed that students will at the same time develop additional grounding in related subjects such as law, economics, planning and construction. We also anticipate that advanced students and valuation practitioners wishing to refresh their knowledge of particular aspects of valuation methodology and practice will find the book useful.

In the limited space available and within the constraints of a single volume, it is not possible to cover everything relevant to the subject and, in any event, making the subject accessible requires a degree of clarity and conciseness so that we have consciously kept chapters short and easy to digest. That is not to say that the subject is simple, some of the chapters are extremely challenging. At the end of most chapters there is some guidance on where to look for further information on a particular topic.

The layout and content of the book represented a considerable challenge not the least in terms of where to start and where to finish. The book is about valuation methodology but it is not about application. We have on a number of occasions within the text referred readers to *Valuation: Principles into Practice,* now in its 6th edition, as the definitive source of guidance on how to put the basic valuation techniques into practice.

Some readers may wish to read the book from cover to cover and we hope there is sufficient narrative logic in the way the chapters are organised and written for this to be a rewarding approach. Others may be interested in certain topics or techniques and may, therefore, wish to focus on particular selected chapters. It may be necessary on occasion to refer backwards and forwards through the text to get a compete picture so, for example, although we deal with the residual method of valuation in Chapter 11, the discounting technique used for calculating the cost of finance is only fully explained later in Chapter 13, which considers investment and compound interest calculations. This is one reason for including the glossary in the early part of the book.

It is our intention to provide coverage of all the valuation techniques that the surveyor is likely to encounter, including methods based on growth explicit discounted cash flow approaches. We do take a fairly traditional approach in that we organise the contents around the five recognised methods of valuation. Our combined experience as teachers of valuation to students new to the discipline leads us to believe that this is the most effective approach and one which accords with general practice. We aim to reflect practice rather than to change it.

The layout of the book

There seems to be no obvious answer to the "where to start?" question but Chapters 1 to 4 provide some preliminary guidance. Chapter 1 examines what is being valued from a legal perspective. The first part of Chapter 2 explores different concepts of value and the second part provides a glossary of terms that have a specific meaning or importance in the context of property valuation. Chapter 3 looks at what the valuer does. This is aimed at providing guidance on developing a set routine around the valuation process. Chapter 3 also considers formal regulatory requirements and how to avoid getting things wrong. Chapter 4 addresses a topic which is rarely if ever considered in valuation texts, the physical inspection. It may be that it is taken for granted that everyone knows how to inspect and measure a property but this is not the case. This is a complex process and practitioners and students will surely find some useful guidance in this chapter.

Chapters 5 to 7 examine the macro and micro-economics of the main property markets with separate consideration of residential and commercial property. These identify among other things the wide range of valuations required by these sectors and some of the more important factors influencing value.

Chapter 8 provides an overview of the five traditional methods of valuation with some useful guidance on how to set out a valuation on the page. This is in our experience another slightly neglected but important aspect of the skill of being a valuer and one which students and some practitioners find quite challenging. Following this, Chapters 9 to 12 provide detailed consideration of the comparison, profits, residual and contractor's methods respectively. Each of these chapters includes examples and practical guidance.

Chapters 13 to 15 examine what tend to be referred to as conventional or traditional methods of investment valuation. In Chapter 13 we consider the time value of money and the fundamentals of compounding and discounting which form the basis of valuation tables. Freehold valuations are explored in Chapter 14 and leaseholds in Chapter 15. In both cases we provide a succession of simple examples which examine how to deal with the most common investment valuation problems that are likely to be encountered in practice. We also provide an exploration of dual rate years purchase methodology notwithstanding the fact that it has gone out of favour in recent years. There a number of reasons for this. First, understanding how dual rate formulae and tables function is a good way of learning about the mechanics of mortgages and other loans and indeed the single rate terminable year's purchase. Second, dual rate years purchase figures may still be encountered. Third, we believe that valuers should be able to choose for themselves from a range of possible methods.

Chapters 16 to 18 are concerned with what might collectively be taken to be contemporary, or growth explicit, approaches. The principles of discounted cash flow are explored in Chapter 16. Here the message is very much that DCF is the same in essence as the investment method of valuation but allows for more sophisticated modelling of income flows and can have a significant role in market analysis. Chapter 17 shows how DCF techniques can be applied to carry out growth explicit valuations and Chapter 18 considers how DCF is used in practice as a tool to support investment analysis.

As a practical guide we wanted to include as many valuation examples as possible. The majority of these have been generated by Excel spreadsheets which have then been rounded manually so that the calculations are internally consistent. It is possible that small rounding errors may occur in some places. As a rule valuation calculations have been made to no more than four decimal places.

The economic context of the book

The book was prepared for publication in the last quarter of 2008. At this point the global economy had suffered its worst financial storm in a century, and many of the world's economies entered a recessionary period that threatened to be deep and prolonged. Many commentators believed that the UK was likely to endure the worst economic downturn it had seen in many decades.

By the end of 2008, the value of stock markets across the globe had fallen dramatically. For example, after a prolonged period of economic growth and increasing asset values, during 2008 the stock exchange FTSE top 100 companies fell in value by around one third, a loss of £465 billion (almost like a bungee jump with someone cutting the elastic). Similarly, commercial property values and residential property values fell, by an average of 30% and 15% respectively. So, although in 2007 the value of the real estate capital markets peaked, 2008 witnessed a sea-change in the global investment environment, following what the International Monetary Fund labelled "the largest financial shock since the great depression" with many of the pillars of the banking credit system crumbling.

Ownership of property is one way of investing money. Periodic property market booms and busts linked to economic cycles illustrate that, although some investors think there is nothing safer than bricks and mortar, property value is not inherent but it is merely a reflection of global, national and local economic health and confidence.

It is essential that those aspiring to be professional valuation surveyors develop the habit of keeping in touch with the way investors think and act. They need to understand what is happening and why in the economy and the world of finance. The financial pages of a good newspaper should be regular reading. Business and money programmes on TV and radio offer clear explanations of the factors affecting change in the economy, including government policy, and the value of investments, particularly property. The *Estates Gazette*, the main weekly UK property magazine, also offers regular readers the opportunity to acquire and maintain a full understanding of the property world.

The features of other forms of investment media, including stocks and shares (equities), bonds and cash in the bank, may be fully explored through other texts referenced in Chapter 18. The glossary in Chapter 2 gives a brief outline of some of main types of investment opportunity. Each category of investment has a different range of key features, which include security of capital and income; liquidity of capital; ease and costs of purchase and sale; divisibility of holdings and cost of investment management. As appropriate, the book compares property investment with key base-line investments, for example "risk free" government gilt-edged securities.

The maths involved in the book

Those new to the study of valuation are sometimes daunted by what they perceive to be a mathematical subject, particularly those who say that "maths has never been my strong subject". Yes, the subject does involve numbers, but there is no need to shy away from it on these grounds. It merely involves simple arithmetic (sums) and occasionally solving equations with an algebraic unknown (where a letter is a shorthand symbol representing a real number). For those lacking confidence in their maths skills internet refresher courses may be useful.

The limitations of the book

The authors accept full responsibility for any errors and apologise for any that have slipped through. The book is not intended to give professional advice and should not be taken as anything other than a general guide to methodology. The authors cannot accept responsibility for the results of any action taken or advice given as a result of reading the book. The addresses of internet sites given in the text were live at the time of writing but inevitably some may cease to operate. Market conditions and the law are as at the date of writing, which is December 2008.

1 Property Ownership

Value is defined in the *Shorter Oxford English Dictionary* as: "The material or monetary worth of a thing: the amount of money, goods etc. for which a thing can be exchanged or traded". It is important to be clear about what is being valued and this chapter focuses on the ownership of *rights* in landed property, or real estate, as these define the thing to be valued. This chapter explores what is meant by real property and examines the different legal interests in property that can exist.

The meaning of real property or land

Property means anything which belongs to a person — that is anything which is proper to that person. Real property (or real estate or realty) refers to one of the two main classes of property, land and buildings. The other main class of property is personal property (or chattels) such as furniture, money and clothing. Property can also be an intangible asset, for example brands, copyrights and patents. These are collectively referred to as intellectual property. This book is only concerned with real property.

Real property or *land* includes not only the surface of the earth, but everything above and below it, from as high up in the sky as is necessary for the ordinary use and enjoyment of the land down to the centre of the earth. Land includes buildings and other structures built on it.

There are different ways of owning, or holding, rights in real property (referred to as tenure, from the Latin *tenere*, meaning to hold) and it is vital to be clear what form of ownership is being valued. It is not possible to own or value property as such because what a land owner has is an *interest* in property, and this interest is what the surveyor is valuing.

This chapter outlines the nature of different forms of property ownership. These provide a framework for the anticipated future benefits associated with the particular form of ownership being valued, for example the right to occupy the property or the possible future rent and capital sums which may be received. Understanding the nature of legal interests is a vital factor in assessing the value of the property concerned.

The nature of legal interests in property

There are two main types of legal interest (which lawyers call "estates") in land, freehold and leasehold.

Freehold interests

All the land in England, Wales and Northern Ireland is owned on a freehold basis. The freeholder (the owner of a freehold interest) is effectively the absolute owner of the land, although, strictly speaking, the land is held under the crown. Some rights to extract minerals are excluded and various authorities have powers to acquire rights over the land, for example to lay pipes or erect electricity pylons, or to compulsorily acquire the interest.

Freehold is the highest form of ownership. The owners of freehold interests have no time-limit imposed on the ownership of their property, so their rights exist *in perpetuity*, that is forever. Freehold owners are able to do what they like with their land, subject to compliance with the law and with any contracts they have entered into with other people.

Freeholders are in a privileged position. They can sell their land if they wish, or, if they retain their property until their death, the freehold interest forms part of their estate and can be passed down to their heirs. Their land ownership may extend to thousands of acres, or they may own a small house, but regardless of the size of their land holding the freeholders will be the kings of their castle. They cannot normally be dispossessed against their will. Probably the only exception to this is where a legal interest is subject to compulsory purchase by local or central government or another body with compulsory purchase powers, in which case they are protected by law from unjustified use of such powers and compensated for their losses.

Companies and other organisations can own interests in land in exactly the same way as individuals. Indeed, most of the central areas of the major towns and cities in the UK are owned by pension funds, local authorities and major property companies.

Freeholders are in a position to exercise a high degree of control over their land. They can choose to develop, improve, maintain or even neglect their property, with no fear that a superior owner will penalise them for their actions. However, freeholders are not entirely free to do what they wish. They are subject to the rights of others and to many statutory restrictions, including the Town and Country Planning Act 1990, which imposes a requirement to obtain planning permission for "development", that is building and similar operations or making a material change of use of the property. This means, for example, the owner of a freehold interest in a property currently used for residential purposes would not be able to start using it as offices without planning permission.

Leasehold interests

Although freeholders are able to sell their freehold interests, normally without hindrance, they may wish to retain ownership and derive a rental income by allowing another to benefit from occupation of the property. In these circumstances the freeholder may create a leasehold interest by granting a lease, which is a legal interest in property which is held either for a fixed time or for a period capable of being fixed by notice.

When a lease is granted the freeholder becomes the landlord (or lessor). The leaseholder is the tenant (or lessee).

The commencement date and duration of a fixed term lease must be certain before the agreement takes effect. While the leasehold interest must be shorter in duration than the freehold interest, it can be created for a short time, for example one year, or a very long time, for example 999 years. Leases do not need to be "in possession", it is possible to grant a lease that starts any time up to 21 years in the future. At common law a lease for a fixed term automatically expires by effluxion (flowing out) of time.

A periodic tenancy (for example a tenancy from year to year, or a quarterly, monthly or weekly tenancy) carries on automatically from period to period until it is terminated by a notice to quit served by one of the parties. The notice required is one full period, or six months, whichever is less.

If a tenant no longer wishes to retain the interest in the property the leasehold may be transferred (or assigned) to a new tenant or surrendered to the landlord.

Ground leases

There are many examples where freehold owners of large sites capable of development choose not to sell their freehold interests to developers, but instead prefer to grant a long leasehold interest of 99 years, or 125 years or more, with leaseholds of 999 years in length not being uncommon. This was common practice in the 19th century when the release of land on a long leasehold basis facilitated the expansion of residential areas in larger towns and cities in the UK. In the modern era this has also happened with freehold land held by local authorities which has been leased to developers on long leases for residential, commercial, industrial and other uses. In these circumstances, the freeholder will receive an annual income from a property developer and their successors (anyone who later buys or inherits the leasehold interest) over the term of the lease. Such an income is normally referred to as a *ground rent*, and the long leasehold interest is commonly referred to as a *ground lease*.

The developer who acquires a ground lease may build on the site. On completion of a residential development the developer may wish to sell the newly constructed houses to owner-occupiers. Since the developer does not possess a freehold interest freehold interests cannot be sold to the owner-occupiers. In such circumstances several new leasehold interests may be granted to each owner-occupier for a term of years which must be less than the term of years the developer holds. The developer would then be referred to as the *head-leaseholder* and would hold the head-leasehold interest. The individual owner-occupiers would be referred to as the *sub-leaseholders* and they would each own a sub-leasehold interest. In this way it is possible for several legal interests to exist in the same property, with a freeholder, a head-leaseholder, and potentially an unlimited number of sub-leaseholders.

Sub-lessees may themselves sublet and thereby create further leasehold interests.

Figure 1.1 The creation of legal interests

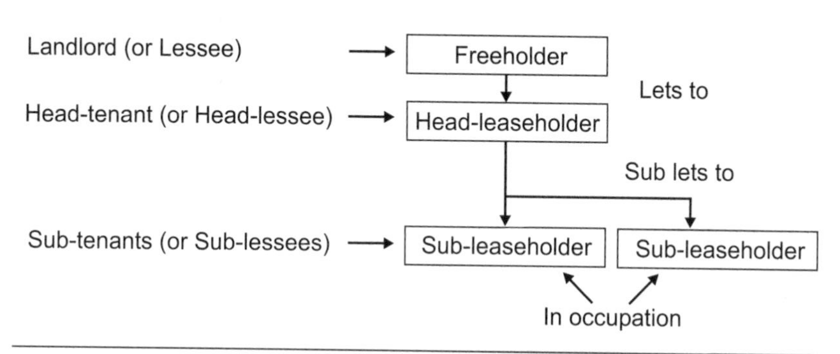

Occupational leases

Leasehold interests are regularly created when the owners of freehold commercial property grant leasehold interests to tenants who wish to occupy property in order to carry on their businesses. These are often referred to as *occupational leases* and may typically be granted for five, 10 or 15 years at the market rent, subject to the leaseholder paying rent quarterly or monthly, in advance. The rent will normally be reviewed at regular intervals, often every five years. Each review is normally an opportunity to increase the rent to the market rent at the time of the review. This of course presupposes that rental values have increased during the period between reviews. Lower quality commercial buildings may be let for much briefer periods, often on yearly, six-monthly or quarterly periodic tenancies.

A major difference between a freehold and a leasehold interest is the degree of control to which a leaseholder is subject. It is clear that the freeholder will only agree to create a leasehold interest if they feel that their own rights will not be harmed by the rights of the leaseholder.

Reversionary interests

A *reversionary interest* is an interest where land is owned by a person but it does not entitle them to present possession. When the term of a leasehold interest expires, the property will revert to the grantor of the interest, who at common law will normally be entitled to occupy it themselves, or to receive its market rent, or to redevelop it.

Statutory regulation, however, may provide exceptions to this rule. Although a detailed examination of these exceptions is beyond the scope of this book they may have a significant impact on the value of the rights of an owner and need to be reflected in the valuation. Examples include the Leasehold Reform Act 1967 and subsequent acts affecting residential property held on long leases and the Landlord and Tenant Acts which give protection to business tenants. For further information on these and other aspects of statutory valuation readers are referred to texts such as *Valuation: Principles into Practice and Statutory Valuation* (see further reading below).

Other legal interests

It is worth mentioning commonhold as an alternative to leasehold interest. This is a relatively new form of tenure which allows freehold ownership of individual flats, houses and non-residential units within a building or an estate. Unlike a leasehold, a commonhold provides a perpetual interest in the individual unit with the rest of the building or estate forming the commonhold being owned and managed jointly by all the individual unit-holders through a commonhold association. The take up of this form of legal interest has not been widespread and it remains fairly unusual.

Other legal interests include easements, such as rights of way and rights to light, wayleaves, allowing access to land or the right to lay and maintain pipes and wires in or over land, restrictive covenants and mortgages.

A restrictive covenant is in effect a promise not to do something. This is specified in the deeds of the property and is binding on successors. It could for example restrict the type and density of development that can be carried out on land, or it could be a promise not to use land to keep chickens or not to build boundary walls around the open-plan front gardens of a housing estate.

A mortgage usually arises in the case of a loan taken out to help finance the purchase of a house or finance the expansion of a business. As security for a loan, a mortgagor (the borrower) grants an interest in their property to the mortgagee (the lender, often a bank or building society) that lends the money. If the mortgagor defaults on the obligation to the mortgagee, the lender may be able to take possession the property in order to sell it and use the proceeds to repay the loan, interest outstanding and costs.

Although it may be quite rare for valuers to come across some of these interests it is clear that the value of a freehold or leasehold interest in land could be affected significantly by their existence.

The importance of leases

A lease is a document that creates a leasehold interest. The lease terms are binding on both the landlord or lessor (the person or organisation granting the lease) and the tenant or lessee (the person or organisation obtaining the leasehold interest).

The lease is a contractual agreement which sets out the terms upon which the tenant will hold the interest, and the leaseholder will then be restricted in their activities on the land by those terms.

Any lease will make reference to the following factors.

- The parties to the agreement — the names of the landlord and the tenant.
- A description of the property, citing, where possible, its address and location, making reference to a plan to show the boundaries of ownership. Any easements, restrictive covenants or other interests affecting the property will also be referred to.
- The start date and length of the lease.
- The amount of rent and the frequency of payment, normally quarterly in advance in the case of commercial property. It is possible that the method of payment may also be specified. The rent will normally be money, but can be in the form of chattels or services. A lease has to have a rent reserved, if the landlord does not wish to collect any rent a rent of one peppercorn (if demanded) may be specified.

In addition, virtually all leases that the valuer is likely to have to deal with include clauses which are tenant's covenants. These are the terms which the tenant agrees to be bound by and will either be positive or negative.

A *positive* covenant will normally commit the tenant to perform a particular task such as to:

- pay the rent in accordance with the terms of the lease
- pay a lump sum or premium
- undertake improvements at the commencement of the term or at some other date

- insure the building, or reimburse the landlord for the cost of insurance
- pay tenant's outgoings, including business rates
- repair the premises, or reimburse the landlord for all or part of the cost of maintenance
- paint and decorate the premises internally and, perhaps, externally at specified intervals
- pay a service charge for services provided by the landlord, particularly relevant when buildings are let to more than one tenant
- yield up possession at the end of the lease.

A *negative* convent prevents the tenant from undertaking certain actions that might harm the interests of the landlord, for example a covenant not to:

- make alterations to the property
- use the premises for any purposes other than the purpose or purposes expressly allowed by the lease (referred to as the user clause)
- assign the leasehold interest without the permission of the landlord. It is normal to add that such consent is not to be unreasonably withheld
- display advertisements without the consent of the landlord.

The lease will also contain a clause with a number of landlord's covenants. For example the landlord may covenant to:

- provide services, carry out repairs, insure the premises and so on
- allow the lessee quiet enjoyment of the property for the duration of the lease.

One of the most important clauses in a modern commercial lease is the rent review clause. This provides for the rent to be reviewed at regular intervals, normally every five years, although periods of three and seven years may occasionally be used and longer review patterns were used in some leases in the past. The rent review clause normally provides that the rent will be reviewed to a market rent, and so allows the landlord to enjoy a rising income over the duration of the lease, provided that the market rent of the property is growing.

The valuer should always check the nature of a leasehold interest in any property which is to be valued, even if is not the leasehold interest that is to be valued. The contents of leases can vary widely, so the ability to read a lease and interpret its provisions in order to assess their effect on the value of an interest in a property is essential. Later chapters will look at the implications of lease terms for the value of property interests.

The result of this legal framework is that there is a market for landlord's interests, usually from property investors, and a separate market for leasehold interests, that is tenant's or occupier's interests. Tenants who have a lease with several years to run, at a rent which is below market rent, will be able to obtain a sum of money for the assignment of their lease.

Summary example

A holds the freehold interest in a vacant city centre development site and grants a building lease of the site to **B** for 125 years at a rent of £100,000pa. The lease provides for a rent review every 5 years to 4% of the market rent of the completed development. **B** builds an office block on the land and lets part to **C** and part to **D**, each paying £2,000,000pa for a 10 year lease, with a rent review to market rent at the fifth year.

Figure 1.2 Relationships between different interests

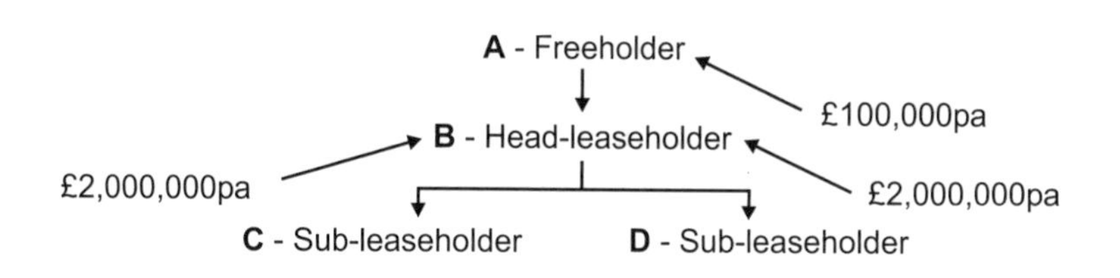

Note

- There are four legal interests in the property and any one can be sold at any time.
- The purchaser of A's freehold interest would receive the ground rent of £100,000pa, and any increased rent following the rent reviews at five yearly intervals for the remainder of the 125 years. At the end of the lease to B, the owner of A's interest would obtain the land and the buildings erected by B

(or any redeveloped buildings on the site). In the meantime, in addition to receiving rent, the purchaser of the freehold would have a measure of control over the use of the property in accordance with the terms of the lease, for example by enforcing user clauses, etc.

- The purchaser of B's head-leasehold interest would be entitled to receive the £2,000,000pa rents from each of C and D, and any increased rent following the five year rent reviews. At the end of the 10 year occupational leases B would expect to re-let the property, on similar occupational leases, generating a continuing flow of (hopefully) growing rental income.
- The purchaser(s) of C and D's sub-leasehold interests would only pay a nominal sum if the rent payable of £2,000,000pa were equal to the market rent. However if, for example, after one year of occupation the market rent of each property were to rise to £2,100,000pa, both C and D would enjoy a rent saving of £100,000pa, that is £2,100,000pa less £2,000,000pa. This 'profit rent' would be enjoyed by purchasers of their interests for each of the next four years, until the first rent review when the rent payable would be increased to market rent (and the profit rent would become zero again). The market value of the sub-leasehold interests would reflect the present value of the right to enjoy profit rents of £100,000pa for the next four years.

The example illustrates how the ownership of the property has been fragmented and the relative value of each legal interest varies depending on the size and duration of their respective future benefits (rent receipts) and obligations (rent payable). The method of valuing these varying legal interests will be examined in later chapters.

This chapter merely gives a basic outline of how land law regulates owners' and occupiers' rights which are relevant to valuation. It explains that a valuer is not merely valuing land and buildings but legal interests in land and buildings. A full knowledge of land law is essential for those who advise developers, owners and occupiers.

Further reading

Law for Estate Management Students, Card, R, Murdoch, J, and Murdoch, S, Butterworths, 5th ed, 1998

Statutory Valuations, Baum, A, et al, EG Books, 4th ed, 2007

Valuation: Principles into Practice, Hayward, R, EG Books, 6th ed, 2007

Concepts of Value and Glossary

2

This chapter aims to do two different things. One is to explain and define some important terms and concepts around value. The other is to provide a glossary of some of the technical terms used in this book.

Other texts may use slightly different definitions and nomenclatures. As far as possible we have tried to use the latest edition of the Red Book (*RICS Valuation Standards*, 6th ed, amended September 2008).

Reading the glossary is about as useful as reading a dictionary, we suggest that you keep referring back to it as you read this book.

CONCEPTS OF VALUE

"When I use a word," Humpty Dumpty said, in a rather scornful tone, "it means just what I choose it to mean — neither more nor less." (*Through the Looking Glass*, Caroll, L, 1871)

In ordinary English usage 'cost', 'price', 'value' and 'worth' are synonymous. But they actually cover a number of different and distinct concepts, and so it is necessary to provide tighter definitions to distinguish the various ideas which the words convey.

Cost

Valuers use the word cost to mean expenditure. This might be the cost of constructing a building or of carrying out repairs or improvements, or the cost of insuring a building, or the fees and other expenses involved in acquiring or selling an interest in property.

Price

Price can be used to mean an asking price, that is the amount at which a seller offers an interest in property for sale, or a sale price — the amount the interest has sold for in the open market.

Value

Value is an estimate of price, and valuation is the art or science of arriving at an opinion of the value of an interest in property.

Because values and costs change with time it is important to decide on the valuation date. In the absence of specific instructions the valuation date is often taken as the date of inspection or the date of the valuation report. Sometimes the valuation date is fixed by statute. The valuation date can never be a date in advance (*RICS Valuation Standards*, 6th ed, Glossary of Terms and UKPS 1.2) because the valuer cannot predict what the market will be like, or even whether the property will still exist, at any point in the future.

The traditional valuation approach works on a current cost, current value assumption. Costs and values are taken as at the valuation date and are not increased to allow for future inflation or growth. More modern methods of valuation can take separate account of future growth potential, these are discussed in Chapter 17.

The *RICS Valuation Standards*, 6th ed, provides a number of bases of valuation, these are shown below. The basis of valuation must be agreed in the terms of engagement and included in a valuation report.

Market Value

"The estimated amount for which a property should exchange on the date of valuation between a willing buyer and a willing seller in an arm's length transaction after proper marketing wherein the parties had each acted knowledgeably, prudently and without compulsion." (PS 3.2, *RICS Valuation Standards*, 6th ed.)

The market value of a legal interest in property is a capital or lump sum. Market Value is probably the most important definition of value. Many valuations are carried out in the context of market transactions and the valuer is often instructed to advise their client on the likely price at which a legal interest in property can be sold.

Market Value is sometimes abbreviated to MV.

Market value replaced an earlier basis of valuation Open Market Value (OMV) in 2004, which had a slightly different definition but the same meaning. The definition was changed to match international standards. OMV, as an expression, is still used in some statutory valuations.

Market Rent

"The estimated amount for which a property, or space within a property, should lease (let) on the date of valuation between a willing lessor and a willing lessee on appropriate lease terms in an arm's length transaction after proper marketing wherein the parties had acted knowledgeably, prudently and without compulsion." (PS 3.4, *RICS Valuation Standards*, 6th ed.)

The Market Rent (MR) is a periodic amount paid by the tenant (or lessee) to the landlord (or lessor). The rent may be payable annually, half yearly, quarterly, monthly or weekly. When preparing a valuation, rents for periods other than a year should be turned to annual amounts.

The market rent is the current rental value of the property at the valuation date. A tenant occupying the property may pay less than the market rent because:

- the lease was granted, or the rent was last reviewed, some time ago and rental values have increased or
- the tenant paid a premium (see glossary below) or
- because the landlord granted the lease at a concessionary rent.

The terms of the lease will affect the market rent, for example a tenant will be prepared to pay more if the landlord covenants to carry out repairs (see Chapter 7 on outgoings), or if there is a longer than normal period before the rent can be increased, or if the rent is paid monthly in advance instead of quarterly in advance. A lease which includes a restrictive user clause, or a break clause in favour of the landlord which limits a tenant's security of tenure, would tend to have a lower market rent.

Older expressions that mean the same as market rent include rack rent, full rental value (FRV) and open market rental value (OMRV).

Capital and income must be kept separate in a valuation to ensure that a capital sum is not accidentally treated as a rental value or vice versa. It is a good idea to put 'pa', or equivalent, after rents to avoid mistaking them for capital values.

Marriage Value

Marriage Value, referred to in the International Valuation Standards as synergistic value, is latent or hidden value released by the merger of two interests. This may be either merging interests in adjoining sites or by merging legal interests in the same property.

Merger of site

Example 2.1

Two adjacent sites, each suitable for a small house, are worth £50,000 each (£100,000 total). Alternatively the site as a whole could be used to build a block of six flats (site value £25,000 each) — value £150,000.

Figure 2.1 Merged site marriage value

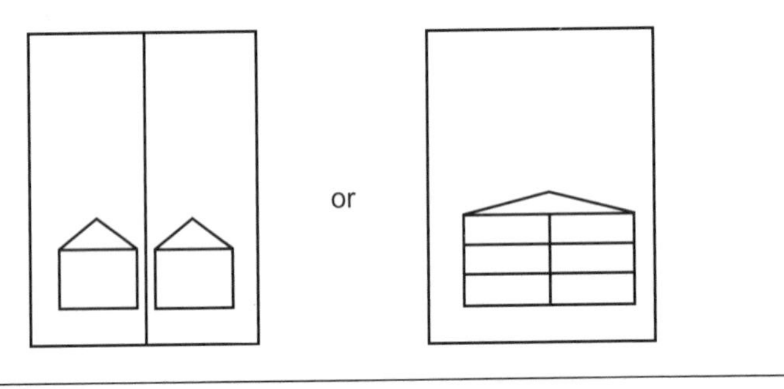

The Marriage Value is the difference between the value of the merged site and the sum of the values of the parts — £150,000 – (£50,000 × 2) = £50,000.

Example 2.2

Two adjacent freehold sites on the edge of a major city have a total area of 12 hectares and are suitable for housing development with a density of 40 houses per hectare. The front site, of two hectares, has access to the highway. The 10 hectares to the rear has no access to the highway. Without proper access it can only be used for agricultural purposes in conjunction with adjoining farm land.

Figure 2.2 Front and back land

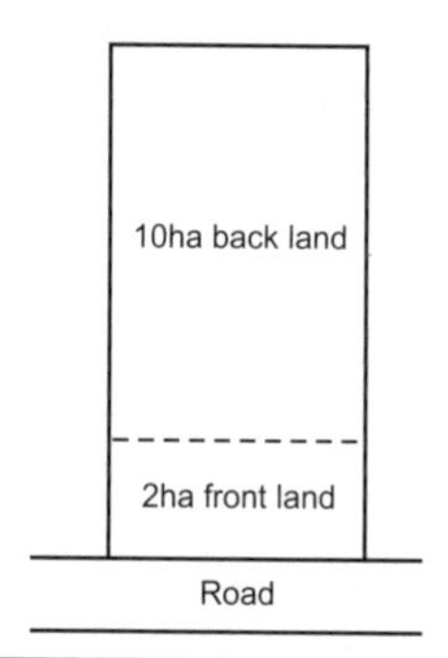

The Marriage Value released on merging the front land and the back land is:

Value of merged interests (12ha @ say £2.5 million/ha)		£30,000,000
Value of front land (2ha @ say £2.5 million/ha)	£5,000,000	
Value of back land (10ha @ say £20,000/ha for agriculture)	£200,000	
		£5,200,000
Marriage Value		£24,800,000

There is potential for ransom strips which provide access to a development site but which will otherwise have a little value.

Merger of legal interests

Example 2.3

The freehold interest in a modern office block let to a tenant has a market value of £1,930,000 (see Chapter 14, Example 14.6). The leasehold interest has a market value of £66,500 (see Chapter 15, Example 15.4). If vacant possession was available the freehold interest would have a market value of £2,000,000.

The Marriage Value released on merging the freehold and leasehold interests is:

Value of merged interests		£2,000,000
Value of freehold interest subject to lease	£1,930,000	
Value of leasehold interest	£66,500	
		£1,996,500
Marriage Value		£3,500

Often the reason for Marriage Value on the merger of interests is the difference in the freehold and leasehold yields and the use of a low accumulative rate in (and possibly tax adjustment of) the sinking fund for the leasehold valuation (see Chapter 15).

Merging a freehold interest with a leasehold interest with a short unexpired term (which is normally the case with modern lease structures) commonly only releases a limited amount of Marriage Value, as shown by Example 2.3. Some property types may show a 'vacant possession premium', if there is considerable demand from owner-occupiers a vacant property will sell for more than an identical property recently let at market rent. This is commonly found in the residential and agricultural markets.

Marriage Value can act as an incentive for owners to sell to each other, or to join in a sale to a third party. It 'oils the wheels'.

The definition of market value excludes the bid of a special purchaser. However this does not mean that Marriage Value should be disregarded if a number of potential buyers in the market would pay 'hope value', that is they will be prepared to bid more in the expectation that the Marriage Value will be obtained at some time in the future.

Fair Value

Where the bid of a special purchaser (see glossary below) is to be taken into account the appropriate basis of value is Fair Value — "The amount for which an asset could be exchanged between knowledgeable, willing parties, in an arm's length transaction" (PS 3.5, *RICS Valuation Standards*, 6th ed, amended September 2008).

Market Value is the minimum a party will accept, and market value plus Marriage Value is the maximum the other party will be prepared to pay. The parties will normally agree a figure somewhere between the two — it is usual to assume a 50/50 split of Marriage Value, but it is a matter for negotiation, for example the Marriage Value could be split in proportion to the value of the interests.

When allocating Marriage Value it is essential to consider the alternatives available to both sides, for example if there are other routes which could be used to access a development site.

Negative Value

It is possible for an interest in property to have a negative value. This can occur if there is a liability which exceeds any value the property may have, for example if the rent paid under a lease is significantly more than the market rent because values have fallen since the lease was granted, or if a leaseholder is responsible for expensive repairs which must be carried out and the lease has only a few years to run.

A negative value only exists if the liability is likely to occur. So, if the cost of erecting a building on a site exceeds the value of the completed development the value of the site for that use is nil because no developer would bid for it. However, the land owner is under no obligation to carry out the development, so the land is not a liability and does not have a negative value. If land was contaminated and the land owner will be forced to carry out remedial works, the cost of which exceed the value of the site after the work is completed, the value of the land will be negative because owning it is a liability.

Worth

Worth is about the performance of property, and can reflect the individual requirements of the client. An estimate of worth can include all the benefits the client expects to obtain from, and the costs which will be incurred in, owning an interest in the property, so the calculation can take into account an investor's own investment criteria, cost of borrowing and tax position.

Different classes of property owner will have different perceptions of worth:

- an investor will see worth as the discounted value of the cash flows from the property
- an owner-occupier running a business will regard the property as a factor of production, and the worth will reflect the contribution the premises make to the enterprise's ability to make a profit. This will include goodwill (see glossary)
- in the public sector worth can include the contribution the property makes to service delivery.

Some of the elements included in worth may be highly subjective and personal to the client. It is difficult to estimate in money terms the contribution that a public library or sports centre makes to the services provided to a community by a local authority.

An estimate of worth may include factors that would not be reflected in the price the property would sell for. A valuer preparing an estimate of worth should highlight such factors in the report and make it clear that the value reported is not the Market Value (see Commentary to PS 3.4, *RICS Valuation Standards*, 6th ed).

Worth can be used to decide if a property should be acquired (the property is worth more to the potential purchaser than the price) or sold (the property is worth less to the owner than the price). When comparing Worth with Market Value it is necessary to adjust Worth for acquisition or disposal costs. Worth can also be used to find pricing anomalies in the market and to justify the retention of property in a portfolio.

In preparing an estimate of Worth it may be necessary to consider if alternative premises or alternative investments would provide a more cost effective solution. For example, imagine a small engineering firm operating from a Victorian building on the fringe of a town centre. The company owns the freehold interest in the premises. The site is ripe for redevelopment and has a Market Value far greater than the value of the firm's profits. An estimate of Worth will show that the property should be sold, but the valuer can go further and consider if there are alternative cheaper premises available which the business could move to, so that the profits can continue to be made and the owner can sell the original property and keep the profitable business running.

Alternative bases of valuation

If a client requests a valuation on some basis other than those specified in the RICS Valuation Standards the basis should be carefully defined and the reason why it is not appropriate to use one of the bases recognised in the Valuation Standards should be explained in the report.

To summarise, price and value are figures which relate to the property market. Cost and worth are not related to the property market.

GLOSSARY

This section is intended to give a brief summary of some of the technical terms used in this book. Words which can be found in a good dictionary have not been included and one-off terms are explained where they occur in the text.

All risks yield the all risks yield is found by dividing the net rent by the capital value of a recently let property. The all risks yield is an explanation for the YP (Years Purchase) used to capitalise the income from a freehold recently let at its market rent — next door sold for 16 times the income, therefore this is worth 16 times the income. The all risks yield is used in traditional investment method valuation to value the reversion. Intuitive adjustments are made to the all risks yield to reflect differences between comparables and the subject property.

Appraisal in the United States appraisal means valuation and valuers are called appraisers. In the past appraisal has been used in the UK to mean an estimate of worth, or a valuation report which included detailed analysis, opinion or advice. Appraisal is now taken to mean the same as valuation (Glossary of Terms, *RICS Valuation Standards*, 6th ed).

Equated yield the internal rate of return which when applied to the projected income flow results in the sum of the incomes discounted at that rate equalling the capital value. Rents at review, lease renewal or re-letting take into account expected future rental growth. Equated yields are used in growth explicit valuations to value income flows which do not have growth potential: see Chapter 17.

Equities are ordinary shares in limited companies. The investor's return is in the form of dividends based on the company's profitability. The stock market is volatile, the value of shares can go down as well as up. There is no guarantee of receiving dividends or of recouping capital. Equity investments are high risk, but they have growth potential and are generally inflation proof because, all being well, the profits (and dividends) will increase and the price of the shares will, long term, keep pace with inflation. Property investments which give a return which has the capacity to grow, such as freehold interests let on short leases or with regular rent reviews are said to be 'equity interests'.

Equivalent yield the internal rate of return which when applied to the projected income flow results in the sum of the incomes discounted at that rate equalling the capital value. Rents at review, lease renewal or re-letting are taken at the current values at the date of valuation. The equivalent yield is used to value all the income flows in equivalent yield valuation (see Chapter 14). As with the all risks yield, intuitive adjustments are made to the equivalent yield to reflect different levels of risk between the comparables and the subject property.

Fixed interest securities are investments which give a fixed return, such as gilts (see below), debentures (loan stock issued by companies), local authority loans (similar to debentures, but less risk as there is little chance of a local authority defaulting) and preference shares (shares issued by companies which pay fixed dividends, preference share dividends are paid before ordinary shares receive a dividend, and any arrears must be met before an ordinary dividend can be paid). Property investments which give a fixed return with no capacity to grow have characteristics similar to fixed interest securities.

Gilts are government stock (that is loans to the government, called gilts because the paper certificates used to have a gold or gilded edge). Gilts may be dated (giving a date when the loan will be repaid or redeemed) or undated (the government can decide when to repay the loan). Gilts are sold in units with a redemption value of £100, the coupon rate is the rate of interest based on the redemption value. The price on the stock market varies (changing the rate of return). The redemption yield includes the return of capital when the stock matures.

Good covenant a tenant who is regarded as low risk, who is unlikely to fail to pay the rent on time or cause management problems, for example a government department, local authority or a well established company with a good reputation. Property let to a good covenant is often regarded as a better investment than similar property let to an ordinary tenant.

Goodwill the price the purchaser of a business will pay in addition to the value of the premises and the stock to reflect the value of the right to receive the profits.

Headline rent the rent taking into account concessions made to the tenant such as rent free periods and reverse premiums. The headline rent is more than the Market Rent. Some rent review clauses require the rent to be increased to the headline rent, although the voluntary Code for Leasing Business Premises in England and Wales 2007 says that Headline Rent review clauses should not be used. Landlords like to quote the headline rents achieved in rent review negotiations and believe that having a rent collected which is more than the market rent increases the value of their interests by more that the cost of the concessions made to the tenant.

Inflation risk free yield the yield which the property investment would show if growth occurred immediately, rather than at periodic rent reviews. The inflation risk free yield is used in some growth explicit valuations, see Chapter 17.

Initial yield also known as the straight yield, the flat yield or the running yield, the initial yield is the current income divided by the capital value. If the property has been recently let the initial yield is the same as the all risks yield, and some authors use initial yield to mean all risks yield.

Lands Tribunal an independent judicial body set up to resolve disputes concerning land. The Tribunal deals with, among other things, rating appeals, compulsory purchase and compensation cases and appeals from Leasehold Valuation Tribunals. For more information see: *http://www.landstribunal.gov.uk.*

Lessee old fashioned legalese for tenant.

Lessor old fashioned legalese for landlord.

Mortgage a loan secured on real property. The lender is the 'mortgagee', the borrower is the 'mortgagor'. With a 'repayment mortgage' the regular payments by the borrower include both interest on the loan and repayment of some of the capital so that at the end of the mortgage the debt has been repaid. Payments on interest only (or bullet) mortgages do not repay any capital leaving the borrower to pay off the debt as a lump sum at the end of the term. This may be done by putting money into an investment fund, for example an endowment policy, an ISA, or a pension. The repayments under a balloon mortgage cover interest and partial repayment with a lump sum payable at the end.

Premium a capital sum paid by the tenant to the landlord in return for the grant of the lease, in exchange for paying a rent less than the market rent. Where a premium is agreed it is normally paid at the start of the lease, although a premium could be paid at some later date. If the tenant has to pay a premium in the future, the amount should be discounted (deferred) and added to the value of the landlord's interest and subtracted from the value of the tenant's interest. A premium which has already been paid must not be included in a valuation. A premium is not, strictly, the price paid for the assignment of a lease, although common usage is to refer to the sale price of a leasehold interest as a premium. Premiums and prices paid for leases are subject to different treatment for Capital Gains Tax.

Prime top quality, used to refer to investment properties, particularly shops. Less good investments are described as secondary or tertiary. The expression 'sub-prime' is now reserved for high risk American mortgages.

Red Book the nickname of the *RICS Valuation Standards*, 6th ed (and its predecessors).

Reverse premium a capital sum paid by the landlord to the tenant in return for accepting a lease. The payment of a reverse premium implies that the rent reserved in the lease is more than the market rent. Reverse premiums are sometimes found in falling property markets where landlords are anxious to preserve the headline rent which they will quote when negotiating rent reviews.

Reverse yield gap where high risk equity investments (see above) show a lower rate of return than "safe" fixed interest securities (such as government stocks), the opposite of what might be expected. The reason the reverse yield gap occurs is that the equities offer a hedge against inflation, while the value of the fixed interest securities is reduced by inflation. A reverse yield gap existed from the early 1960s until inflation was brought under control in the 1990s.

RICS the Royal Institution of Chartered Surveyors. The RICS is the leading professional body in the UK, with members in many other countries world wide. For further information see *www.rics.org*.

Special assumption assumptions are suppositions that are taken to be true. Some assumptions are listed in the Red Book as special assumptions, for example an assumption that planning permission has been granted, that building works have been completed, or that property is vacant when it is occupied or vice versa. Special assumptions may only be made if they are realistic. Where it is necessary to make special assumptions in order to prepare a valuation, the special assumptions must be agreed with the client in writing. If a client requests a valuation on the basis of a special assumption that the member considers unrealistic, the instruction should be declined (PS 2.2, *RICS Valuation Standards*, 6th ed). A valuation report must set out any special assumptions made, and include a statement that they have been agreed with the client (PS 6.4, *RICS Valuation Standards*, 6th ed).

Special purchaser someone who has a particular reason to want to purchase a particular interest in a property, and will therefore be prepared to pay more than Market Value. A special purchaser might already own another interest in the property, or an interest in adjoining property, and so be able to bid more than other potential buyers because marriage value will be released if the interests are merged.

Years Purchase A multiplier applied to an annual income to find its capital value. It represents the present value of each £1pa of income for a given number of years discounted at a given rate of compound interest.

Yield the rate of return on an investment. A number of different types of yield are used in property valuation. See all risks yield, equated yield, equivalent yield, initial yield and inflation risk free yield.

Further reading

Property and Money, Brett, M, 2nd ed, Estates Gazette, 1997
RICS Valuation Standards, 6th ed, 2007, amended September 2008, particularly PS 3
The Glossary of Property Terms, Jones Lang LaSalle, EG Books, 2004

3 The Role of the Valuer

Valuation is generally regarded as a hybrid discipline sitting somewhere between science and art. There is a range of valuation methodologies that can be applied to solve particular valuation problems and some of these appear to involve quite complex looking mathematics and formulae. Add to this the fact that valuation has its roots in the discipline of economics and what we have looks like a science. However, many students of valuation, especially those with a science background, do sometimes get confused and frustrated by its lack of precision and its reliance on judgment and opinion and, in this respect, valuation can take on the characteristics of an art. Of course, if valuation was all precision and no judgment, a simple matter of applying a formula, it is unlikely that the skills of the valuer would be much in demand and we would all be out of a job. So, this hybridism is perhaps something we should celebrate.

This chapter considers in outline what the valuer does as well as how that role is regulated by professional standards and codes of conduct. It then goes on to consider briefly those occasions when a valuation is challenged and what happens if the valuer falls short of those standards. Finally, the chapter explores how to develop a logical and disciplined approach to valuations which should help to minimise the risk of a claim for professional negligence.

Demand for and purposes of valuation

The work of the valuer can be divided into two broad categories; market valuations and hypothetical valuations. Market valuations are those that will be tested in the market. For example, when a valuer takes an instruction to value a house for sale purposes, the opinion expressed and reported to the client will be tested by reference to the sale price ultimately achieved. Hypothetical valuations, on the other hand, are usually opinions of value that will not normally be tested in the market. So, a valuer, instructed to value the same house for inheritance tax purposes, may have to defend the opinion of value in negotiation and possibly even in court, but it is unlikely to be tested in the market. The distinction is not always clear cut. Take for example a loan valuation. Although this might be seen as a hypothetical valuation as no immediate sale is envisaged, problems may arise at some later date when, following the borrower defaulting on the loan repayments, the lending institution repossesses the property and tries to realise the value of the security by a sale on the open market.

Figure 3.1 shows the range of types of valuation that the valuer might be instructed to undertake. For example, a valuer might be required to undertake valuations for insurance, development and redevelopment, for balance sheet purposes and rating and other forms of taxation. Agency valuations will normally be undertaken on the basis of market value or market rent for sales or lettings whereas statutory valuations, such as valuations for taxation and compulsory purchase, will mostly be subject to specific definitions governed by particular acts of parliament and case law. All valuation activity, though, will be governed by professional codes of conduct and the majority of these valuations will also be regulated by national and international standards.

Professional body regulation and codes of conduct

Standards in valuation are set by valuation guidance, such as the RICS Red Book. This provides the regulatory framework within which valuations are carried out. With increasing globalisation, international standards are becoming more widely recognised and a regulatory framework is now emerging with local regulations sitting within these international standards.

Figure 3.1 Types of Valuation

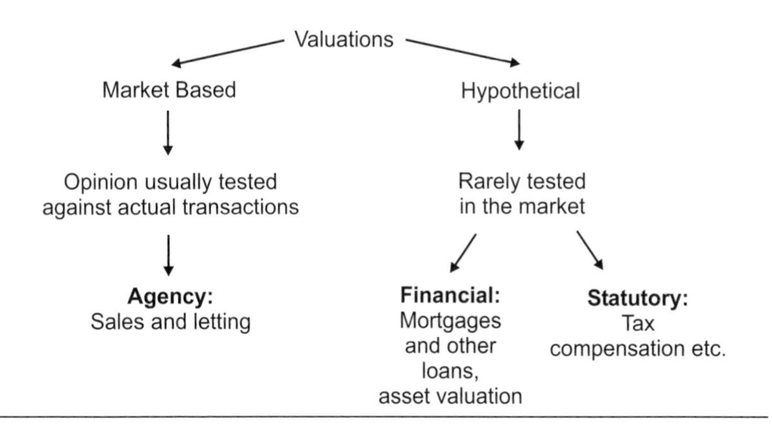

The current RICS Code of Conduct was introduced with effect from 2007. This contains simple rules dealing with the regulation of members which set out "the standards of professional conduct and practice expected of Members of the RICS". In addition to these rules, members are expected to follow all general legal obligations. The rules concern the following broad areas of professional practice:

- integrity and conflicts of interest
- competence
- service and customer care
- lifelong learning
- solvency
- provision of information to the RICS
- co-operation with the RICS.

Members breaching the rules may be subject to disciplinary proceedings as set out in the RICS bye-laws. Both the International Valuation Standards Committee (IVSC) and the European Group of Valuers' Associations (TEGoVA) subscribe to similar ethical codes. IVSC statements are included within the International Valuation Standards and TEGoVA publishes the so called Blue Book (equivalent to the RICS Red Book). In all cases the rules are common-sense sets of ethical standards to be expected of any professional person.

The evolution of national and international standards

Bodies such as the RICS, the International Valuation Standards Committee (IVSC), the Appraisers Association of America who publish The Uniform Standards of Professional Appraisal Practice (USPAP) (North America) and TEGoVA (Europe) are responsible for the regulation of a wide range of valuations undertaken within their own jurisdictions.

RICS regulation is through what is fondly known as the Red Book, now in its 6th edition. To understand the evolution of the Red Book it is necessary to go back to the trauma of the property crash of the early 1970s. Then, as more recently in 2007–08, the property crash was accompanied by a crisis in the banking sector. During the boom years of 1971–73 financial institutions had been lending on the back of a dramatically rising property market that everyone at the time seemed to think would carry on indefinitely. The boom did come to an end and the RICS Assets Valuation Standards Committee (AVSC, now the Assets Valuation Standards Board) published its first guidance note in 1974, coinciding with the property crash. In 1976 the RICS published the first edition of the Red Book, *Guidance Notes on the Valuation of Assets*. The Red Book only applied to asset valuations incorporated in published accounts, and was voluntary.

A 'White Book', the *Manual of Valuation Guidance Notes* was published in 1980 to give guidance on other valuations. A second edition of the Red Book, also called *Guidance Notes on the Valuation of Assets*, was published in 1981, and a third edition, with a new title, *Statements of Asset Valuation Practice and Guidance Notes*, appeared in 1990.

A summary of the evolution of the Red Book from 1970 to 2007 is set out below.

1970s	Property market collapse
1976	1st ed *Guidance Notes on the Valuation of Assets*
1980	White Book
1981	Red Book 2nd ed
1990	3rd ed *Statements of Asset Valuation Practice and Guidance Notes*
1991	Red Book becomes mandatory for asset valuations
1994	Mallinson Report
1996	4th ed RICS *Appraisal and Valuation Manual* incorporating the White Book. The Red Book became mandatory for most valuations.

2002 The Carsburg Report
2003 5th ed RICS *Appraisal and Valuation Standards*
2007 6th ed RICS *Valuation Standards*.

From 1991 valuers were required to comply with the AVSC Statements of Asset Valuation Practice and Guidance Notes, making the Red Book mandatory for valuations of assets for incorporation into company accounts and other financial statements, or referred to in published or public documents, and for investment or security purposes. Where there were special circumstances making it inappropriate to comply with the Red Book the valuer could make a clear statement to that effect in the valuation certificate, giving the reasons for the departure.

Following the market crash in the early 1990s a committee was set up under Michael Mallinson to "Investigate, comment upon and produce recommendations on any aspect of commercial property valuation".

Mallinson's principal concerns about many valuations centred upon their reliability, credibility and clarity as well as the regulation of those carrying out valuations. The Mallinson Report, published in 1994, identified minimum standards to be established and the need to understand and satisfy the needs of the client. Mallinson's recommendation was the introduction of "a comprehensive and user-friendly manual of valuation standards and guidance". In response to the Mallinson Report, the Red Book was rewritten, incorporating the White Book, and the new 4th edition was published in 1996 as the *RICS Appraisal and Valuation Manual*. At the time some commentators, tongue in cheek, suggested that the merged AVSC manual and the White Book should be known as the Pink Book or even the Red and White striped book, but common sense was allowed to prevail. The new 4th edition of the Red Book applied to and was mandatory for virtually all valuations.

Their have been subsequent reports on the relationship between valuers and their clients, including the Carsburg Report (2002) which made a number of recommendations concerning the accuracy and currency of valuations, independence and objectivity, valuation reporting and quality assurance and monitoring.

The 5th edition, *RICS Appraisal and Valuation Standards*, was published in 2003. Much of the detailed guidance on valuation methods had been removed (it was about half the size of the fourth edition), and the 2004 amendments integrated International Valuation Standards, including replacing Open Market Value with Market Value.

The 6th ed, *RICS Valuation Standards*, was published in 2007 and came into effect in January 2008. The 6th edition incorporates the RICS Rules of Conduct which were introduced in 2007.

One of the ways in which the standards have evolved over time is an increasing harmonisation between the Red Book and the IVSC's International Valuation Standards. This is the inevitable consequence of an increasingly global property market and although there are some tensions between local and international practice, harmonisation is likely to continue. It is expected that the TEGoVA will follow suit with its own version reflecting wider global practice.

The basic structure of the current Red Book is set out below.

Introduction
Glossary

Practice Statements

PS 1	Compliance and ethical requirements
PS 2	Agreement of terms of engagement
PS 3	Basis of value
PS 4	Applications
PS 5	Investigations
PS 6	Valuation reports

Guidance Notes

GN 1	Trade related property valuations (inc goodwill)
GN 2	Plant and equipment
GN 3	Valuation of portfolios and groups of properties
GN 4	Mineral bearing land and waste management sites
GN 5	Valuation uncertainty

UK Practice Statements

UKPS 1	Valuations for financial statements
UKPS 2	Valuation for financial statements — specific applications
UKPS 3	Valuation for loan facilities
UKPS 4	Residential property valuation (other than for mortgage)
UKPS 5	Regulated purpose valuations (valuations prepared under PS 4.1 or UKPS 2)

UK Guidance Notes

UK GN 1	Inspections and material considerations
UK GN 2	Shared ownership of residential property
UK GN 3	Valuations for capital gains tax, inheritance tax and stamp duty
UK GN 4	Valuations for charities
UK GN 5	Local authority disposal of land at less than best consideration

Practice statements are individual statements comprising a short statement or rule. This is followed by a commentary giving additional information to assist in interpretation and application of that rule and appendices containing supporting information. Practice Statements 1–6 are general statements that apply to all states (state is defined as an organised political entity and in most cases equates to country).

Guidance Notes are statements of good valuation practice and procedure covering areas such as Inspections and Material Considerations (UK GN 1) and Valuations for Charities (UK GN 4).

It is important to understand that the Red Book was concerned with the mechanics of valuation practice and not valuation techniques. The RICS publishes Valuation Information Papers which are intended to provide information, outline current practice and give an indication of the approach to issues that may arise in the subjects to which they relate. Examples include Valuation of Owner-Occupied Property for Financial Statements (No. 1) and The Capital and Rental Valuation of Hotels in the UK (No. 6).

The Red Book applies to all valuations where there is a possibility that third parties (that is anyone other than the client or the valuer) might rely on a valuation report. This could include loan valuations or valuations of assets that are likely to appear in company accounts or elsewhere in the public domain. A shrinking number of types of valuation remain exempt. These include advice during the course of litigation; the decisions and reports of arbitrators, independent experts and mediators and advice given during negotiation; internal valuations; certain agency and brokerage work and the valuation of antiques and fine art.

Valuation accuracy and the courts

Both the RICS Rules of Conduct and the Red Book contain a clear expectation that professional valuers will exercise due skill, care and diligence. It is important to explore, briefly, just what this means in the event of a valuation being challenged on the grounds that the valuer has fallen short of these expectations, and stands accused of professional negligence. In a useful paper exploring the approaches of the courts in negligence cases against valuers, Crosby (2000) identified three ways of considering how good a valuation actually is. These are accuracy, variation and bias. Accuracy is the extent to which a valuation is similar to a subsequent transaction in the open market. Variation is the extent to which different valuers will arrive at the same or similar conclusions as to their opinion of value. Bias is the extent to which valuations are consistently under or above the ultimate sale price. (This last test is obviously rather difficult to apply to non-market valuations.)

During a rising or stable market general increases in value will tend to mask valuer error and, in any event, losses are unlikely in such circumstances. But, every so often, a recession in the property market will lead to challenges, claims of negligence and the intervention of the courts. The principal defence against a claim for negligence is that the valuer acted with the care and diligence expected of a professional person working in valuation. However, because valuation is subjective, and needs to take account of a large number of variables, judgments about the duty of care owed by the valuer to a client are anything but straight forward.

In the past the courts have tended to focus on the result of the valuation as well as the process undertaken to arrive at that result. A valuation could be shown to be incorrect but it is unlikely that the courts will become involved unless this has resulted in a loss. Even then the valuer might be able to resist such a claim on the grounds that the result of the valuation was the result of correct application of valuation methodology and that other professional valuers might easily have arrived at the same conclusion. This position is summarised below.

	Right process	**Wrong process**
Right outcome	Clearly there can be no claim for negligence	The valuer has arrived at the right conclusion, albeit by luck rather than judgment, and will probably escape a claim for negligence because it is unlikely that any loss can be proven
Wrong outcome	The valuer could be negligent but may be able to defend a claim on the grounds of sound process, especially if it can be demonstrated that others would have reached similar conclusions	In these circumstances the valuer would have no defence and, in the event of a claim would almost certainly be considered to be negligent

In many cases the courts have applied a margin of error test to the outcome of the valuation. While it can be dangerous to treat such cases as precedents, it would seem that a 10–15% error either side of the valuation figure may well be seen as acceptable. This is often referred to as bracketing. What is clear is that the courts are unlikely to expect pinpoint accuracy.

The valuation process

Process is defined in the *Shorter Oxford Dictionary* as: "a course of action or a procedure, especially a series of stages in manufacture or some other operation". Process is rather important in valuation. It may be the only defence that stands between the valuer and a successful claim for negligence. Certainly if things do go wrong with a valuation, it is very useful to be able to demonstrate that everything possible was done to arrive at a sound conclusion in terms of the opinion of value expressed.

The prospect of undertaking a valuation should be an exciting challenge and, human nature being what it is, there is a temptation to rush into the task and reach a conclusion as quickly as possible. However, valuations are complex assignments with many different elements. There is always a risk that, if too hasty a conclusion is reached, something important might be missed. As with any complex task, it is a good idea to break the valuation down into smaller stages to be tackled more or less sequentially. The process recommended here consists of four distinct stages, each defined by a number of questions to be posed by the valuer. These stages and questions are set out in the table below. It is recommended that inexperienced valuers follow this process (or something very similar). In addition, the fact that such a process has been followed should be carefully detailed in the case file and retained as a record. After a while, this disciplined and logical approach will become second nature, but even then it should never be taken for granted.

Stage	Questions
Preliminaries	Who is the client, what does the client need and what service is to be provided? What is the valuation for? What is being valued?

Market and economic factors	What is the state of the market? What evidence do you have and how can this be analysed? Will there be any special purchasers?
Methodology	What valuation method(s) should be used? What calculations might be needed? How do you weigh up all the information?
Conclusions	How are you going to report your opinion to the client?

The process should be seen as broadly, but not rigidly, chronological. Clearly it would be inappropriate to report an opinion of value to the client before agreeing terms of engagement, just as it would be pointless to apply a particular valuation methodology before identifying the property to be valued. This is not to say that parts of the process are not iterative or can even be carried out in parallel. So, market research and analysis of transactions may well be ongoing throughout the process.

Each of the stages and questions is now considered in turn.

Preliminaries

1. What does the client need and what service is to be provided?

These two questions are at the centre of the need for regulation of the valuation process. The valuer has to understand the client's needs and identify the precise level of service that will be provided. Overlooking this simple step has been one of the major causes of professional negligence claims against valuers. The Red Book, in particular, was a response to a lack of transparency in this area and requires that the terms of engagement are confirmed to the client in writing before the valuation report is delivered. In practice, it is common (and prudent) to get the client's agreement to the terms of engagement rather than simply writing to confirm what they are. This ensures that both parties are clear about the precise nature of the instruction, gives the client an opportunity to raise questions and means that, if the letter is "lost in the post", the surveyor is not vulnerable to a client who denies all knowledge of the terms when a problem arises.

The minimum terms of engagement are set out in Practice Statement 2 (PS2). A list of these minimum terms is shown below and readers are referred to PS2 for a detailed explanation of each item:

(a) identification of the client
(b) purpose of valuation
(c) subject of the valuation
(d) interest to be valued (ie the tenure of the subject property)
(e) type and classification of property
(f) basis of value (see Chapter 2)
(g) date of valuation
(h) disclosure of material involvement
(i) status of the valuer
(j) currency to be adopted
(k) assumptions or departures
(l) extent of investigations
(m) nature and source of information
(n) consent to and restriction on publication
(o) limits of exclusion of liability
(p) valuation to be undertaken in accordance with the Red Book
(q) fee basis
(r) complaints handling procedure
(s) monitoring under RICS conduct and disciplinary regulations.

Where a valuer acts for a client on a regular basis it is acceptable to have standing terms of engagement which the parties agree will apply to all the valuations for that client.

2. *What is the valuation for?*

The purpose of the valuation needs to be agreed as part of the instruction. This is an essential pre-requisite as the purpose of the valuation will almost certainly determine whether or not the instruction is covered by the Red Book and may well be significant in determining the valuation methods to be adopted. It will also determine the valuation basis and any specific statutory definitions of value. In certain cases the purpose of the valuation may prescribe the valuation date.

3. *What is being valued?*

The identification of what is being valued is of course critical. This question needs to be considered from both a physical and a legal perspective. Gathering factual information about the property is often taken for granted but care needs to be taken that nothing is overlooked.

This will usually involve a physical inspection to identify and record the extent of the property, including measurements, as well as the physical state and condition. It is also a good idea to check boundaries so that these can be confirmed with the client. Physical inspections should also be seen as a logical process and readers are referred to Chapter 4 for a detailed consideration of the inspection of the property.

What is being valued is not so much the physical entity itself but a bundle of legal rights in, on and over the land. Legal rights cannot be seen although some, such as the existence of a tenant, may be physically manifest. The tenure of the property, details of any tenancies, rights of way, wayleaves, easements and restrictive convents all need to be considered as any of these could have a significant impact on the value of any property. For further information on property rights please refer to Chapter 1.

Market and economic factors

4. What is the state of the market?

Value is an economic concept. In all circumstances value arises from the balance between supply and demand (see Chapter 5). This varies over time and will vary with property location, type and class. Property markets can be highly discriminating so that, even in a recession, certain sub-types might be the subject of high demand. Valuers are expected to have a very sound understanding of macro and micro-economic forces within their own areas of expertise and this may extend to cover local, regional, national or even international market conditions depending upon the type of property being valued.

5. What evidence do you have and how can this be analysed?

Where possible the majority of valuations will be based on market evidence. There are some exceptions but one of the key roles of the valuer will be to interpret what is happening in various property markets, through the collection and analysis of market data. This will depend upon the valuer's own records although, increasingly, information relating to market activity and transactions is available from a range of electronic sources. Whatever the source of evidence, care should be taken to gain a full understanding of the circumstances surrounding individual transactions, especially anything that might have influenced one or more of the parties to the transaction. For instance, the transaction may not be at arm's length, or the purchaser or vendor might have been under extreme pressure to buy or sell. These and many other things should be part of the valuer's market intelligence. All data should be treated with scepticism and carefully analysed and assessed for accuracy and reliability.

6. Will there be any special purchasers?

With market valuations, value is often generated by so called special purchasers. One of the more subtle skills required of the valuer is the ability to identify the existence of one or more potential purchasers with a particular motivation for acquiring a given property. This is important as the special purchaser can often generate an overbid producing surprising results. In some hypothetical valuations the bid of a special purchaser has been excluded. This has been the cause of considerable controversy in areas such as compulsory purchase and capital taxation and, in general terms, both types of valuation would tend now to include the special purchaser's bid in most cases.

Special purchasers may be adjoining owners who perceive some additional benefit will accrue by adding further property to an existing holding. They might be purchasers with special circumstances such as a particular tax position or they might be owners with another interest in the same property. In all cases the valuer needs to be aware of who might be in the market and why and how this might influence any bid. For a further discussion of the concept of special purchaser readers are referred to Chapter 2.

Methodology

7. What valuation method(s) should be used?

Valuation textbooks generally recognise five distinct methods of valuation: comparison, investment, profits, residual and contractor's (see Chapters 8 to 13). In some cases it might be appropriate to use more than one method, for example where the subject of the valuation contains different types of buildings. Or it might be sensible in some cases to use more than one method as a check on the main valuation. This might happen with development land (residual and comparison methods) and some business properties (profits and comparison methods).

8. What calculations might be needed?

Not all valuations require complex mathematical calculations but many do. This is true of most investment valuations but would also be the case with residual valuations and with valuations carried out using the contractor's method. Profits method valuations might require a detailed analysis of business accounts. But even the comparison method is likely to involve some numerical analysis of comparable data. In complex cases, this may even include statistical analysis of data such as changes in value over time. Much of this, though, is routine and common sense, although where a valuation is likely to involve large volumes of data, consideration should be given to the need to use specialist software or spreadsheet applications.

9. How do you weigh up all the information?

This in essence is what the art and science of valuation is all about. Much of the foregoing is important but routine, and the well informed and well organised valuer should have no problems getting to this point. However, if you have explored the first eight questions set out above with care and diligence, you will now be in possession of a large volume of information about the subject of your valuation, no matter how small and insignificant the property might actually be. How then do you manage all the information you have gathered, decide on its relative importance and use all of this to generate an opinion of value?

All the evidence needs to be weighed, in other words it needs to be assessed in terms of its relevance and its validity. For example, evidence of transactions closer to the valuation date of the subject property will normally be given more weight than evidence of more distant transactions. Once this is done firm conclusions should begin to emerge.

Conclusions

10. How are you going to report your opinion to the client?

The conclusion of any valuation is the communication of an opinion of value. Whether or not this is tested in the market, it will normally be communicated to the client. This may be an oral transmission but it is more likely to be in the form of a written report. Many of these reports will be covered by the Red Book and, even where this is not the case, the Red Book provides a useful template covering the sort of things that should be included (PS 6.1). These can be summarised as follows:

(a) identification of the client
(b) the purpose of the valuation
(c) the subject of the valuation
(d) the interest to be valued
(e) the type of property and how it is used
(f) the basis, or bases, of the valuation (to be stated in full PS 6.2)
(g) the date of valuation
(h) disclosure by the valuer of any material involvement or a statement that there has not been any previous material involvement
(i) if required, a statement of the status of the valuer (internal or external)
(j) the currency that has been adopted, where appropriate
(k) any assumptions, special assumptions, reservations, any special instructions or departures
(l) the extent of the surveyor's investigations
(m) the nature and source of information relied on by the valuer
(n) any consent to, or restrictions on, publication
(o) any limits or exclusion of liability to parties other than the client
(p) confirmation that the valuation accords with the *RICS Valuation Standards*, 6th ed
(q) a statement of the valuation approach
(r) the opinion(s) of value in figures and words
(s) signature and date of the report.

Obviously, the valuation report is written with the client in mind. However, it is sometimes referred to as "the silent witness" and it should, among other things, provide a clear record of the way in which the valuer has followed a logical and thorough process in undertaking the valuation, and should provide good contemporary evidence that the valuer has demonstrated a duty of care. What should be clear, above all, is the relationship between the terms of engagement and the report and the way in which these two documents link with the valuation process.

In cases of negligence valuers are required to demonstrate that they are not in breach of their duty of care to their client (and to third parties who may have relied on the report). Professionals are judged on the generally accepted practice of their profession, that of an ordinary competent person working in valuation.

This can be determined by looking at a margin for error either side of what is seen as the correct valuation. However, this depends on a subjective judgment and it may well be sufficient for the valuer to demonstrate that the correct process and methodology have been followed in undertaking the valuation.

Compliance with the Red Book requires minimum terms of engagement to be agreed with the client. It also specifies the minimum content of valuation reports. Both reflect the importance of a logical and sequential valuation process as an essential pre-requisite for "reliability, credibility and clarity" of valuations.

Finally, the habitual adoption of a set process in undertaking all valuations will help to ensure that nothing is missed. This may be even more important for the inexperienced valuer who may find valuation tasks complicated and intimidating at first or, worst still, be tempted to jump to conclusions before considering all the facts and all the available evidence.

Further reading

RICS Valuations Standards, 6th ed, 2007. Practice Statements 2 and 6 provide more detailed guidance on terms of engagement and the content of valuation reports.

RICS Regulation *Rules of Conduct for Members* (2007).

"Valuation accuracy, variation and bias in the context of standards and expectations", Crosby, N *Journal of Property Investment & Finance*, (2000) 18 (2) pp 130–161. The first part of Crosby's paper provides a useful review of key cases on valuation negligence and the balance between the need for valuation accuracy and sound procedures.

4 Property Inspection

A valuation almost always involves an inspection of the property which is the subject of the instructions and its immediate vicinity. In some circumstances an inspection is not required, for example:

- if a revaluation is requested and the surveyor is satisfied that there have been no material changes to the property or its location since the last inspection or
- if the client requires a desk top or drive by valuation or
- if an automated valuation model (that is a computer based system which analyses sales data to provide an estimate of value without any input from a surveyor) is to be used.

The terms of engagement must indicate if the property is not to be inspected and appropriate assumptions must be included in the report.

The surveyor needs to have a clear understanding of where the premises are and what land and buildings are included. While an address and post code will often be adequate, some properties do not have clear boundaries and this can lead to some of the property being excluded from the valuation or someone else's property being accidentally included, with the potential for a negligence claim. For example a freehold house may occasionally be found which has part of the garden held on a periodic tenancy with no obvious boundaries visible on site. In *Platform Funding Ltd* v *Bank of Scotland plc* [2008] 42 EG 168 a surveyor, instructed to prepare a report for mortgage purposes on a house under construction which was little more than a shell, with no windows or roof, was misled by the borrower into inspecting a nearby property which had almost been completed. The Court of Appeal held that a surveyor has an absolute obligation to inspect and value the correct property, even if reasonable care had been taken. It is, therefore, prudent for the valuer to ensure that the property inspected is adequately described in the report, possibly with the aid of a plan and/or reference to the area of the site and/or the buildings, and that the client should be advised to have the legal title to the property checked against the details given in the report before any transaction is completed.

RICS Valuation Standards, 6th ed, Practice Statement 5.1 says "Inspections and investigations must always be carried out to the extent necessary to produce a valuation which is professionally adequate for its purpose". The commentary to PS 5.1 goes on to note "In settling the terms of engagement the valuer must agree the extent to which the subject property(ies) are to be inspected and the extent of any investigations to be made. Where a property is inspected the degree of on-site investigation that is appropriate will vary, depending upon the nature of the property, the purpose of the valuation and the terms of engagement agreed with the client".

So the terms of engagement (discussed in Chapter 3) should set out the extent of the inspection which will be carried out, and, perhaps more importantly, should make clear what the valuer will not do. The inspection will not normally be a building survey (formerly known as a structural survey, a name which was felt to be misleading as it led to expectations that a surveyor would report on the whole structure of a building whether it could be examined or not), and it is usual to specify that:

- the valuer will not carry out a building survey and will not inspect any part of the property which is covered, unexposed or inaccessible, which will be assumed to be in good repair and condition. The valuer cannot express an opinion about or give advice upon the condition of parts of the property which are not inspected and the valuation should not be taken as making any implied representation or statement about such parts
- the surveyor will not move any obstruction including furniture and floor coverings.

The valuer should not make an assumption that a potential problem does not exist if it would be apparent from routine enquiries or an inspection of the property and its vicinity. If an assumption is agreed in the terms of engagement that a problem does not exist but it is evident on inspection that the assumption is, or is likely to be, wrong, the valuer should discuss the issue with the client and agree revised assumptions before the report is issued.

Where the limited inspection suggests that a problem may exist the surveyor must "follow the trail" in so far as is possible. In *Smith* v *Eric S Bush (a firm)* [1989]1 EGLR 169 a valuer noticed that the first floor chimney breasts in a house had been removed, but failed to check that the chimney stack was adequately supported. Eighteen months later one of the chimneys collapsed and fell through the bedroom ceiling. The House of Lords held that the valuer was negligent.

This does not mean that valuers are obliged to extend an inspection significantly beyond the terms of engagement, or beyond their expertise. Surveyors must beware of giving an opinion on an issue if they do not have appropriate training or skill. If necessary the matter can be raised with the client, with a view to either obtaining a more extensive investigation, specialist advice, and if necessary an estimate of the cost of remedial works, or making a special assumption, depending on the purpose of the valuation and the level of risk involved.

In *Beaton* v *Nationwide Building Society* [1991] 2 EGLR 145 the judge said that the duty of a valuer inspecting a house for a building society was "not to carry out a structural survey, but, although his inspection was more limited, it required the exercise of the skill of a reasonably competent professional valuer. ... It should have been evident to him from such indications as the visible stepped fractures, from the existence of the clay subsoil with shallow foundations and from the oak trees growing at a distance less than their own height from the house, that there was a risk of subsidence causing continual structural movement. He did not attach sufficient importance to the signs. The advice to the purchasers should have been that no mortgage advance could properly be made on the property unless and until a full investigation had been carried out into the causes of the structural movement and the remedial action required".

Material considerations

Material considerations are matters which may have an impact on the market's perception of the value of the property. The policies of lenders and insurers will have a significant effect on demand. Many potential buyers require a loan secured on the property before they can complete a purchase, and if a mortgage isn't available they will not be able to buy the property. Lenders normally insist on their security being insured, and most property owners will want insurance against a range of risks, so not being able to obtain insurance cover on the usual terms will again restrict the value of the property.

RICS Valuation Standards, 6th ed, UK Guidance Note 1 says "Unless expressly stated to the contrary in the terms of engagement ... Valuers have an obligation to investigate, consider and report on any material feature that affects a property or its surroundings that could impact on value ...".

Many material considerations will be apparent during an inspection provided that the surveyor knows what to look for. Inspecting the property also gives the surveyor the opportunity to verify the information relied on in preparing the valuation and check, in so far as is possible, that the assumptions agreed in the terms of engagement are valid.

Location, location, location

The location of the property is a major factor affecting value. Such things as social and commercial facilities (including the availability of jobs/labour, the presence of shopping facilities and car parking), the proximity to transport infrastructure (roads, rail, bus and tram services and airports) and the nature, use and state of repair of neighbouring properties will all have an impact on demand, and hence value.

For shops the position in the street or shopping centre may have an effect on value. Being close to a magnet (a shop which draws customers, like a supermarket), the pedestrian exit to a car park or a street crossing will result in increased foot fall and more passing trade, while adjoining dead frontage like banks and offices is a disadvantage. Restricted on-street parking or vehicular access can reduce values significantly.

Location is less crucial for offices, although car parking and public transport will often be important.

The value of residential property can vary significantly over a very short distance. Being in the catchment area of good schools, having pleasant views, a south facing aspect and a neighbourhood or post code with a good reputation will all enhance value. Negative factors can include being in an undesirable district, or on part of an estate with a dubious reputation, proximity to busy roads, aircraft flight paths, and bad neighbours (consider, for example, living next to a fast food shop, a pub car park or a sewage works).

Some locations are affected by environmental factors which can have an effect on the value of property, for example:

- areas where mining has been carried out (with the risk of subsidence, which can occur many years after mining has ceased; where appropriate the surveyor should recommend that a mining report be obtained)
- the presence of radon gas (associated with granite, when instructed to value property in affected areas the surveyor should recommend that testing should take place if the results of a test are not available)
- methane (from nearby landfill sites)
- flooding and coastal erosion (risks which are increasing because of climate change, development on floodplains and changes in government policy on coastal defence) and
- electromagnetic radiation (a perception of risks associated with nearby high voltage electricity lines and sub-stations, satellite masts and mobile phone transmitters may affect marketability).

The potential presence of these factors can be established by inspection, routine enquiries or local knowledge. The terms of engagement should specify the limitations on any investigations and the assumptions that will be made.

Proposals for future development nearby can affect the value of property, which in extreme cases can leave it blighted. Surveyors will not normally make enquiries of the local authority to ascertain if there are any adverse proposals likely to affect the property, and will assume that there are no planning or highways proposals that are likely to have an effect on the value of the premises unless these are specifically notified to them.

The property itself

The use of the land and buildings is often the reason why there is demand for it, and hence value. It is normal to assume that the existing use is lawful and that planning consent and building regulations approval have been obtained for all development on the site.

If the property is situated in a conservation area, or includes listed buildings, or if any trees on the site are protected by Tree Preservation

Orders, maintenance costs may be higher than normal and development potential may be restricted.

The buildings

Buildings on the site will often represent a considerable proportion of the value of the whole. Their construction, age and anticipated future life will have a significant effect on the value of the property.

Buildings may include potentially hazardous materials. Asbestos, in common use in buildings until the mid-1980s, can be expensive to maintain and remove. The surveyor should examine the management plan which is required for non-domestic property under the Control of Asbestos Regulations 2006. Other hazardous materials include lead, some composite panels, and timber after some chemical treatments.

Some buildings are constructed with deleterious materials. These are materials which degrade with age and cause structural problems. Examples include:

- high alumina cement (used in concrete beams until the mid-1970s some of which failed causing buildings to collapse, the risk being particularly high with beams over 5m long. However the mere presence of any high alumina cement may prevent a loan being available and therefore blight a sale)
- Mundic (concrete made with sulphide bearing mine wastes, used in the South West from about 1920 to the 1950s)
- calcium chloride additives (used to speed the setting of concrete, but which may cause corrosion of reinforcing rods)
- black ash mortar (increasing the risk of cavity wall tie failure)
- flat roof coverings (always vulnerable to leaks, although designs have improved considerably) and
- red shale (used under solid floors, which expands if wet, causing structural movement).

There are limits to the extent of the valuer's investigations and this is clearly set out in the Red Book: "A valuer will not normally be competent to advise on either the nature, or risks, of contamination or hazardous substances, or any costs involved with their removal. However, where valuers have prior knowledge of the locality and experience of the type of property being valued, they can reasonably be expected to comment on the potential that may exist for contamination, and the impact this could have on value and marketability. It will therefore be necessary for the valuer to state the limits on the investigations that will be undertaken and any sources of information or assumptions that will be relied upon." (*RICS Valuation Standards*, 6th ed, Appendix 2.2 Assumptions).

For example: "no investigation will be carried out to determine whether or not any deleterious or hazardous material has been used in the construction of the property, or has since been incorporated, and the surveyor is, therefore, unable to report that the property is free from risk in this respect. The surveyor will assume that such investigation would not disclose the presence of any such material".

Caution should be exercised when valuing non-traditional forms of construction, particularly pre-cast reinforced concrete (PRC) houses. The following types are listed in the Housing Act 1985 as defective: Airey, Boot, Boswell, Cornish, Dorran, Dyke, Gregory, Myton, Newland, Orlit, Parkinson, Reema Hollow Panel, Schindler and Hawksley SGS, Stent, Stonecrete, Tarran, Underdown, Unity and Butterley, Waller, Wates, Wessex, Winget, Wollaway and Smith (BSC). In *Peach* v *Iain G Chalmers* [1992] 2 EGLR 135 a house described by a surveyor as "concrete block built, harled" (harling is a type of rendering used in Scotland) was in fact of Dorran construction. It was held that the construction of the property was an essential element in the valuation and the defendants fell below the acceptable standard in failing to note that the house was of PRC construction. The failure to identify the type of house was an act of negligence.

Some PRC house types not on the defective list have inherent structural problems. In *Izzard* v *Field Palmer (a firm)* [2000] 1 EGLR 177 a surveyor provided a mortgage valuation report on a maisonette which noted that the property was built to the Jesperson industrial system but did not provide any warnings. Part of the building failed, the owners were unable to sell their property, which was repossessed and eventually sold at a considerable loss. The Court of Appeal held that the surveyor was negligent in failing to warn about the structure of the building, the recommendations of the Building Research Establishment in relation to that type of construction and that fact that the service charge could increase.

In describing property, estate agents will stress selling features which will attract a purchaser. The valuer should consider all the factors which will affect value, including whether the buildings are functional and adaptable, and of good or bad architecture and design. For domestic property the type (terrace, semi-detached or detached, and whether a house, bungalow, flat or maisonette) will have an effect on value. For shops and industrial property the ability to load and unload goods easily and the absence of steps and ramps is important.

The dimensions and floor areas of the buildings will often be needed to prepare a valuation. A rough guide to the RICS Code of Measuring Practice appears at the end of this chapter. Purchasers and tenants of non-domestic property commonly pay a price per square metre, and while adopting a price per unit area is rarely appropriate for domestic property, at least in the UK, a large house will normally be worth more than a smaller one. For industrial premises ceiling heights and floor loading will be significant. With shops, frontage is more important than depth (see Chapter 9 on zoning). For most types of property upper floors will be less valuable than ground floor space unless, for flats and prestige offices, the top floor can reasonably be described as a penthouse.

The services, installations, fixtures and fittings and any plant and machinery which form an integral part of the building need to be considered. In addition to electricity, gas, water and drainage a building may have telephone and IT cables, heating and air conditioning systems, lifts, stair lifts, escalators, fire detection and alarm systems, security alarms, CCTV systems, window cleaning equipment, electrically operated doors, entry phone systems, swimming pools and other leisure facilities. The inspection and testing of these services and equipment is usually outside the scope of valuation instructions, and they, and any associated controls or software, will normally be assumed to be in working order and free from defect.

The state of repair and condition of the premises will affect their value. Minor disrepair which does not materially affect value is not normally reported, and the terms of engagement should reflect this. It is usual to assume that buildings and other structures on the site are in good repair except for defects which are specifically noted, and that all reasonable internal and external repairs and maintenance have been carried out.

Exceptional risks, such as flooding, subsidence, landslip or settlement, abnormal risk of impact, proximity to inflammable liquids and the use of flammable materials in the structure, are factors likely to affect the availability of insurance or the level of premiums, which will have an effect on marketability.

An Energy Performance Certificate is required when a building is constructed, sold or rented out. In January 2008 CBI/GVA Grimley's Corporate Real Estate Survey reported that 61% of companies would pay more rent for a green building although 81% of these would only pay marginally more. It may be that rising fuel costs will make energy efficiency a more important factor.

The Disability Discrimination Act 1995 (as amended by the Disability Discrimination Act 2005), imposes a duty on the occupiers of premises to allow disabled people to make use of the services provided. This can often be achieved by adapting the services rather than making changes to the building, however poor access may limit value.

The land

The dimensions and area of the site, whether the amount of land is appropriate for the buildings, the adequacy of access, whether adjoining highways are adopted and maintainable at the public expense, whether the site slopes and any inconvenient changes in level will all affect potential purchasers' or tenants' perception of the property.

The condition of boundary fences and walls is often overlooked by the market, however repair or replacement can be costly, especially where retaining walls are involved.

Vegetation on the site is also a factor that potential purchasers and tenants do not pay particular attention to, despite the risk that trees can cause movement to buildings (especially on clay soils) and the cost of managing invasive species such as Japanese Knotweed.

It is common to limit the inspection to the general state of the site and only comment on matters which are likely to materially affect the value of the property.

Many urban areas have had a number of previous uses which may have contaminated the land with potentially hazardous or harmful substances, for example: heavy metals, oils, fuel, solvents, poisons and pollutants. These contaminants require expensive specialist treatment, and appropriate assumptions need to be stated in the valuation report, for example — "the surveyor is not aware of the content of any environmental audit, land quality statement or other environmental investigation or soil survey which may have been carried out on the property and which may draw attention to any contamination or to the possibility of any such contamination. The surveyor will assume that no contaminative or potentially contaminative uses have ever been carried out on the property. The surveyor will not carry out any investigation into past or present uses either of the property or of any neighbouring land to establish whether there is any potential for contamination from these sites to the subject property and will therefore assume that none exists. Should it be established subsequently that contamination exists at the property or on any neighbouring land or that the property has been or is being put to a contaminative use this might reduce the value reported".

Exclusion clauses designed to safeguard the surveyor from a negligence claim do not necessarily give the protection hoped for. The Unfair Contract Terms Act 1977 applies to the terms of engagement, and a surveyor could be liable, despite the statement in the paragraph above, if the courts found it to be unreasonable given the ready availability of historic land use data on line.

Where specialist reports, including an estimate of any costs of remediation and the impact on marketability, are available the valuation can be prepared on the assumption that the information provided is correct.

Steeply sloping sites may have been levelled, and development may have taken place on land which has been quarried and then back filled. Such land may be unstable, causing structural problems with buildings.

Property with development potential

The availability of planning consent is crucial to the value of property with development potential. If planning permission has not been obtained the valuer has to make a realistic special assumption about what consent will be available, including any conditions that would be imposed. In *Montlake* v *Lambert Smith Hampton* [2004] 3 EGLR 149 a valuer disregarded the development potential of a rugby pitch. It was held that a non-negligent valuation should have drawn attention to real planning prospects.

Physical restrictions on development, including geotechnical conditions, plants and animals protected by the Wildlife and Countryside Act 1981, and the presence of archaeological remains can all significantly restrict development value, and it should be recommended that any assumptions that these factors do not exist should be verified by appropriate specialist reports before purchase.

Legal factors

The legal interest in the property to be valued, if leasehold or let the unexpired term of the lease, the rent and any rent review clause, and whether vacant possession is available, are crucial to a valuation. In the absence of a current detailed report on the property's legal title by the client's lawyers, the valuer must state what information has been relied on and what assumptions have been made about the tenure of the property, including easements, wayleaves, covenants and lettings (see Chapter 1).

A number of recent fraud cases have involved attempts to mislead valuers by providing copies of false leases showing inflated rents. The surveyor should, therefore, check that details of rents receivable are in line with the market for property of that type in the area and that any other information provided rings true.

It is common to assume that there are no encumbrances on title and that leases do not contain any onerous or unusual provisions which would affect value.

There may be a difference between the value of a property with vacant possession and the same property occupied by leaseholders, even if the tenants are paying the market rent. This will apply particularly to property with development potential. The ability of a freeholder to obtain vacant possession of let property may be limited by statute, tenants of business premises will often have security of tenure under the Landlord and Tenant Act 1954, long leaseholders of houses have rights to purchase the freehold and extend their leases under the Leasehold Reform Act 1967, and owners of flats often have the right to extend their lease or collectively buy the freehold under the Leasehold Reform, Housing and Urban Development Act 1993. Any assumption that the property is vacant when it is occupied, or vice versa, is a special assumption (see Chapter 2, Glossary).

In England and Wales individual flats should be leasehold or commonhold — a freehold flat and any property with a flying freehold (that is where freehold property in different ownership is above or below the subject property) is not mortgageable because the repairing covenants are not enforceable.

It is prudent to recommend that assumptions relating to title (including details of the property inspected), the existence of planning consents and building regulations approval, and the results of local searches and mining searches (if appropriate) should be verified by the client's lawyer before any transaction is completed.

The inspection

Having agreed the terms of engagement, and therefore the extent of the inspection, there are a number of things to sort out before turning up at the property.

The client should be encouraged to provide as much information as possible to help the surveyor and to support the assumptions being made, including details of any relevant agreements, approvals, consents, notices, reports, tests and plans which are available.

Arrangements need to be made for access. If the property is occupied an appointment should be arranged, and enquiries made about any activities on the premises which might put the surveyor at risk. If the property is vacant keys may be required, and it is sensible to inquire if any buildings have been boarded up and if any security alarms have been fitted. The surveyor should also ask if the property is in a derelict or dangerous condition. Health and safety is an over-riding issue, and a surveyor should always be accompanied if there is any risk involved.

In all cases the surveyor should ensure that someone knows when and where the inspection is taking place and that the alarm will be raised if they do not return on time.

The Health and Safety at Work Act 1974 imposes a duty on employers to provide and maintain equipment and systems of work that are safe and without risk to the health of employees or others who may be affected by their undertaking. Employees are required to take reasonable care of their own safety and that of others who may be affected by their acts or omissions. Surveyors should ensure that they are familiar with their organisation's health and safety policy and arrangements for implementing safe working procedures and that they are aware of any hazards which may exist, together with any safe working instructions issued by clients (abstracted from "About Surveying Safely" *http://www.rics.org/Practiceareas/Management/Healthandsafety* accessed 24 July 2008).

Before leaving the office the surveyor should check that all the necessary equipment is available and in working order. This invariably includes paper, pens and pencils (or a hand held dictating machine), a tape or other measuring device, and an ID card or some business cards. Depending on the extent of the instructions the surveyor may also require such things as a ladder, a damp meter, a torch, binoculars, a probe and a camera. Other items which may be needed include a street plan, road map (and/or SatNav), waterproof and protective clothing, wellington boots, a hard hat, safety boots, spare batteries for anything that needs them and a mobile telephone. The latter should not be relied on as a safety device, some areas have poor signal reception and it is possible to fall through a roof or floor in a way that prevents access to the pocket or pouch containing the mobile phone.

The surveyor must make and retain notes of the findings and, in particular, the limits of the inspection, the circumstances in which it was carried out, and who provided information, especially if it cannot be verified.

Site notes are vital if there is a dispute. If the notes have been dictated the tape should be typed up. Hand written notes should be legible and in a form to facilitate writing the report, although the surveyor should not attempt to draft the report while the inspection is being carried out as this prevents adequate reflection on the findings. *Watts* v *Morrow* [1991] 2 EGLR 152 concerned an allegation of negligence against a building surveyor. The judge said "It is the practice of the defendant to dictate his survey as he walks around the property during his inspection. He does not dictate notes into a dictating machine, he dictates his survey report into a dictating machine on site. When he returns to the office, he gives the tapes to his secretary, who types them up and the report is then amended and sent to the client. That was the practice adopted by the defendant on this occasion, so that he had no notes to disclose on discovery of documents. It also led to his report being lengthy and diffuse and to its conclusion being inadequate ... It led to a report which was strong on immediate detail but excessively, and I regretfully have to say negligently, weak on reflective thought".

There is no correct method for a property inspection, each surveyor will develop their own preferred order and format depending on the nature of the inspection, the type of property and their own inspection approach. Whatever format is chosen changes to the order of inspection should be avoided to minimise the risk of missing something.

The rest of this section offers guidance on carrying out a simple inspection of a basic property.

The first step, on arriving at the address, is to start filling in the field sheet with the address, date, weather conditions, the surveyor's and any assistant's initials, the client's name and any contacts' names. Then have a look before approaching the property, think health and safety, consider whether anything might drop on or otherwise injure you and watch where you are putting your feet.

Meet the occupier, explain who you are, what you are there for and get their consent to what you want to do. Check that any health and safety issues are dealt with and note any areas where access is not available. Some surveyors prefer the occupier or their representative to come round with them. It is helpful to be on good terms with the occupier, who can often provide useful information.

When inspecting the interior of any buildings follow a systematic approach, visiting each floor in turn, and move in one direction (clockwise or anti-clockwise), noting the use of each room or circulation area. Cellars, roof spaces and under floor voids (if included in the inspection) are often left to the end to minimise the amount of mess caused.

Use a fixed routine for each room or area, take any measurements required and note the condition of the ceiling, walls (have load bearing walls or chimney breasts been removed?), fixtures and fittings, doors and window frames (check that they square and use a probe to test for signs of rot) and decorations, look for any signs of dampness and use the moisture meter (if required). Examine the floors where exposed, decide if they are level, are timber floors springy? Note the presence of all services (in country properties do not assume that gas and drainage are connected to the mains, cunningly hidden propane tanks and cess pits or septic tanks can easily mislead the unwary). Beware of lead cold water pipes and DIY electrical installations.

Outside note the construction of each building, the materials and any defects in a systematic way, considering each elevation in turn, looking at the roof and associated features such as chimney stacks and flashings, the walls (look particularly for any structural movement, the presence of tie bars should never be disregarded), the rain water goods and soil and vent pipe, the external joinery (again use a probe to test for signs of rot), the damp proof course (if present check the height above ground level and that it is not bridged), and consider sub-floor ventilation for suspended timber floors. Having examined the building at close range, stand back, and take an overall view of each elevation. The objective is to form a general opinion of the whole property, not just produce a list of defects. Bear in mind the need to follow the trail, just saying that there is a damp patch to a ceiling or a raking crack to the brickwork above a window is not particularly helpful.

Walk around the site, paying particular attention to boundaries, trees, access arrangements and any signs of problems with drains. Take any outside measurements required.

Look out for flying freeholds — repairing covenants may not be enforceable and the property may not be accepted as suitable security for a loan, which will have a significant effect on value.

Note any rights which are required for the benefit of the subject property and if the property is subject to any third party rights (unexpected occupiers might be trespassers or hold leases or tenancies which were not included in the assumptions on title), including public rights of way. These factors should all be included in the report.

Make a note of any factors limiting the inspection, for example areas which were inaccessible.

Look at neighbouring property (particularly noting the use and any signs of disrepair), then the locality generally. Note any agents' for sale or to let boards which will give an indication of any comparables on the market and which firms are operating in the area.

It is always worth making a note of anything unusual about the inspection. While the green parrot in the back bedroom is irrelevant to the job, it might help the surveyor to remember the property months later, and in any event will give the impression that it can be remembered ("oh yes, that was the property with the green parrot in the back bedroom").

Finally, before leaving the site check that all the information required has been collected. It is always embarrassing to have to go back!

Code of Measuring Practice

The RICS Code of Measuring Practice (Code of Measuring Practice, RICS Guidance Note, 6th ed, 2007) is a widely accepted industry standard which gives precise definitions to allow accurate measurement of buildings and land on a common and consistent basis. The Code, which provides bases of measurement and gives detailed guidance on when each should be used, runs to 42 pages including helpful explanatory diagrams.

The first stage of any valuation negotiation between two surveyors is often to discuss the areas of the buildings or land they are dealing with. Their initial figures will rarely be the same. The Code of Measuring Practice means that they should be measuring the same thing, so it is just a question of agreeing what the dimensions are, there will always be minor variations in measurements taken on site. Adopting the wrong basis of measurement can have serious consequences. In *United Bank of Kuwait* v *Prudential Property Services Ltd*

[1994] 2 EGLR 100 among the reasons for a negligent valuation was the incorrect use of the gross internal area instead of the net internal area.

It is only possible to give a rough guide to the RICS Code of Measuring Practice here. There are three main bases of measurement.

GEA Gross External Area the area of a building measured externally at each floor level, excluding canopies, external open sided balconies and roof terraces. The GEA includes garages, conservatories and outbuildings which are attached to the property, but not greenhouses, garden stores, fuel stores and the like in residential property. GEA is used for residential insurance valuations and for planning applications.

GIA Gross Internal Area the area of a building measured internally to the outside walls at each floor level. The GIA includes garages, conservatories and outbuildings which are attached to the property, but not greenhouses, garden stores, fuel stores and the like in residential property. GIA is used for non-residential building cost estimation, and for the valuation of industrial buildings, warehouses, department stores, variety stores, food superstores and leisure property. GIA could mislead purchasers and tenants and so the basis of measurement should be stated when it is used.

NIA Net Internal Area the usable area within a building measured to the internal face of the perimeter walls at each floor level, including kitchens but excluding among other things bathrooms, toilets, internal structural walls, areas with a headroom of less than 1.5m, cleaners' rooms, stair and lift wells, lift rooms and permanent (not notional) lobbies and common areas. NIA is used for the valuation of shops, supermarkets, offices and any business uses not measured to GIA.

There is no accepted practice for measuring residential property for valuation purposes. Estate agents commonly describe the accommodation using linear measurements (usually omitting the hall, landing, staircase, bathrooms and toilets) and the GIA is often quoted for new build property.

The basis of measurement used should be stated in the valuer's report.

For land measurement the *Site Area* is the total area of the site within the site boundaries, measured in a horizontal plane. *Gross Site Area* is the site area, plus half roads adjoining, enclosed by extending the side boundaries of the site, up to the centre of the road, or to 6m out from the frontage, whichever is less. Gross Site Area is mainly used for industrial property and warehouses, however the area of land quoted in old deeds for all types of property sometimes include half road widths adjoining — another trap for the unwary!

The Code of Measuring Practice recommends that metric units should be adopted. Square metres are used for buildings and small sites, larger sites are measured in hectares (10,000 m^2). Imperial units are still in widespread use by surveyors. The following factors allow conversion between the two systems:

1m = 39.3701 ins, 3.28084 ft, 1.09361 yds
$1m^2$ = 10.7639 sq ft, 1.19599 sq yds.
1 ha = 2.47105 acres or 11,959.9 sq yds.
1 in = 2.54cm or 0.0254m.
1 ft = 0.3048, 1 sq ft = $0.0929m^2$.
1 yd = 0.9144m, 1 sq yd = $0.836m^2$.
1 acre = 4,840 sq yds = 0.404686 ha.

Occasionally more archaic imperial units may be found in old deeds:

A chain is 22 yds (the length of a cricket pitch and 1/10th of a furlong).
A rod, pole or perch is 5 1/2 yards (1/4 chain)
A square rod, pole or perch is 30 1/4 sq yds or 1/160th of an acre.
A rood is 40 square rods, poles or perches, or 1/4 acre.

Further reading

Building Surveys and Inspections of Commercial and Industrial Property, RICS Guidance Note, 3rd ed, 2005.
Building Surveys of Residential Property, RICS Guidance Note, 2nd ed, 2004.
Contamination and Environmental Matters — their implications for property professionals, RICS Guidance Note, 2nd ed, 2003.
Code of Measuring Practice, RICS Guidance Note, 6th ed, 2007.
RICS Valuation Standards, 6th ed, 2007, particularly PS 5 and UK GN 1.

5 Property Markets and Economics

Valuation is generally regarded as a branch of the discipline of land economics. At its very simplest level, economics is concerned with the way in which demand and supply interact to generate value, and the valuer needs to have a detailed understanding of the factors which influence the supply of and demand for different types of property. It is important to note at the outset that this chapter refers to *markets* in the plural, because part of the complexity of the underlying economics of property stems from the fact that we are dealing not with one but with many different markets and sub-markets that can be differentiated in terms of both sector and location. It can even be argued that markets can be separated in terms of the purpose for which property is acquired.

This chapter looks at the demand and supply sides of the property markets, exploring some of the many factors which impact upon demand and supply, how these factors change over time, as well as some of the imperfections of the property markets and how they impact upon value.

Markets

There are a number of ways of thinking about markets. Possibly the obvious image that comes to mind is a physical space where buyers and sellers can gather together to buy and sell fruit, vegetables, fish and other commodities. Alternatively, we might conjure up something like the London Stock Exchange, again a physical space but where buyers and sellers don't necessarily gather but at least communicate through brokers to make deals over stocks and shares and commodities. Within the property context, we might consider the auction house where buyers gather together to bid competitively for what is on offer. Finally, markets can now also exist as virtual entities, such as eBay, where buyers and sellers connect with one another via the internet.

These are all very different types of market, but they do have some common features and it is these features that make them markets. In economics it is possible to identify something referred to as the 'perfect market'. The perfect market will exhibit the following characteristics:

- the goods on sale are all the same — the product is homogenous
- there will be a large number of buyers and sellers so that no individual will be able to exert undue influence on the market
- those buyers and sellers will have perfect information about prices, available products and alternatives
- it is assumed that the buyers and sellers will all act rationally to achieve the best bargain from their perspective
- the quantity of goods sold in any transaction will be small in relation to the quantity available, so that a single deal leaves the market unaffected
- and finally, there will be perfect mobility so that buyers and sellers are free to enter and leave the market at will.

These so called perfect markets are of course idealised constructs. Sometimes stock markets are regarded as the nearest thing to a perfect market but even they fall short of the ideal requirements. It should be obvious by now that property markets are even further from this ideal and it is clear that economists see property markets as imperfect markets.

Property markets

There are at least four distinct ways of looking at the plurality of property markets; by property type or sector, by location, by type of demand and by quality. These are illustrated in the table on p61.

This is illustrative rather than exhaustive. The sector column is fairly self explanatory and the main property market sectors should be familiar enough and are normally seen as distinct markets with different factors driving demand and supply. So, for example, the

Sector	Location	Type of demand	Quality
Residential Retail Office Industrial Leisure Agricultural	Local Regional National International	Occupation Ownership Speculation Investment Development	Prime Secondary Tertiary

agricultural market often bucks the trend in other markets because it acts as a good hedge (that is protection against possible loss) in times of recession when other sectors are under pressure.

Within each main sector it might also be possible to identify any number of sub-markets and each of these may well be quite distinct. So, for example, it might be possible to distinguish between specialised industrial and light industrial or within the retail sector between lock up shops and retail warehouses. Each sub-sector is likely to be subject to its own particular influences which in turn impact upon supply and demand.

The second variable is location. The residential market for owner-occupation is generally regarded as fairly parochial in that its physical extent is very locally based and limited; a post code district or primary school catchment area for example. At the other extreme, the market for large prestige HQ office buildings for international corporations may transcend national boundaries.

It might also be possible to subdivide markets in terms of the purpose of acquisition. There may be a distinction between property acquired for ownership and property acquired for occupation and even property which is acquired for the purpose of speculation; that is the hope of making a short term gain, as distinct from investment where property is held for gain but over a longer period of time. A further distinction may be made in terms of property acquired specifically for development or redevelopment purposes.

A final means of distinction is concerned with the quality of the property. Arguably this applies to the investment market more than the other markets. Prime property usually refers to the best property in the best locations, occupied by high quality tenants ("good covenants") that are well-established companies having stable earnings and no excessive liabilities. Such tenants are sometimes referred to as Blue Chip tenants. (The source of this phrase is thought to be the game of poker, where the colour blue denotes the highest value chip.)

The term 'prime' then would apply, for example, to city centre shops occupied by strong national multiples or to modern office blocks occupied by government departments. Such properties will usually attract the lowest investment yields, in other words their market values, that is their capital values, will be high relative to their market rents. Distinct markets will also exist for secondary and tertiary locations and poorer tenant covenants. Secondary and tertiary properties will have higher yields to reflect higher levels of risk and lower levels of potential rental and capital growth.

The significant point here is that each of these markets will be subject to different pressures influencing supply and demand, so that they will tend to behave differently and each sub-market will require a degree of specialist understanding on the part of the valuer.

Demand and supply and elasticity

The price of most commodities is a function of the relationship between supply and demand. Two successive hot summers in 1976 and 1977 had a detrimental effect on the national potato crop, the supply of potatoes went down and the price went up. Even the cost of a bag of chips was higher as a result! On the other side of the equation, the emergence of China and India as industrial powers in the early part of the 21st century increased the demand for a whole range of commodities and was one of the drivers behind a tripling of crude oil prices and significant increases in the price of metals. This relationship between supply and demand is relatively easy to understand in principle, although comprehending the underlying reasons for changes in supply and demand, in markets as complex as the different property sectors, is rather more challenging. Analysing property market demand and supply is very much the job of the valuer who must understand the many factors affecting these two sides of the price equation. Each will be considered in turn. But first it is necessary to consider the principle of elasticity.

Figures 5.1 to 5.3 provide a simple illustration of the theoretical relationship between supply and demand, but readers wishing to explore this in more detail should refer to the further reading at the end of this chapter.

The basic forces of supply and demand are probably best illustrated by reference to a simple commodity such as brussels sprouts. The demand curve is a downward slope which predicts that if the price of sprouts rises the quantity demanded will fall as shown by the demand curve D_1. At price P_1 the quantity demanded will be Q_1.

Figure 5.1 Demand

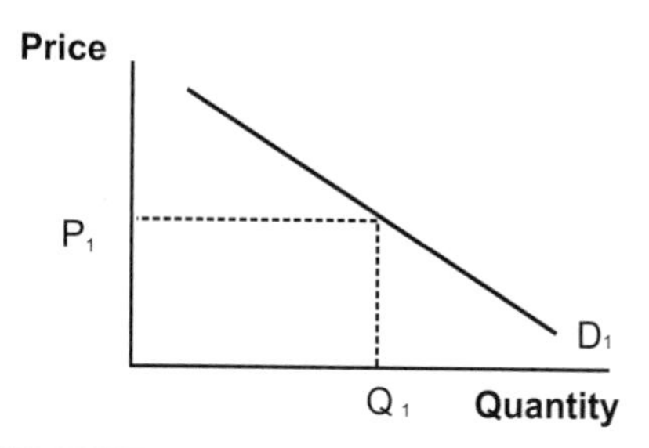

The supply curve by contrast is upward sloping predicting that at higher prices, higher quantities will be will be supplied. This is because high prices will encourage new producers to enter the market.

Figure 5.2 Supply

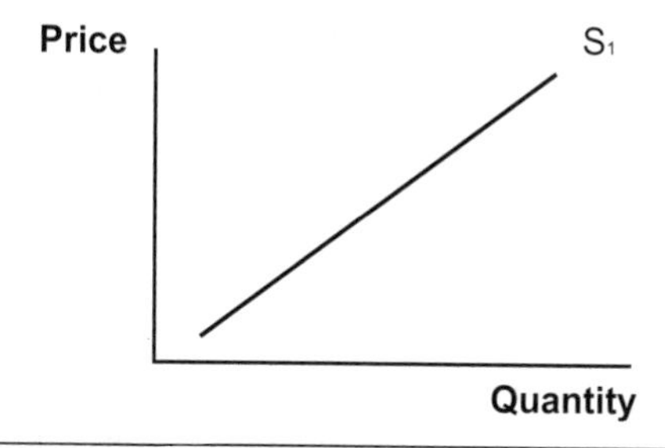

Demand and supply theory suggests that demand and supply will tend towards an equilibrium point where the demand and supply curves intersect. At this point the amount of brussels sprouts demanded by consumers is equal to the amount producers are prepared and able to supply.

Figure 5.3 Supply, demand and equilibrium

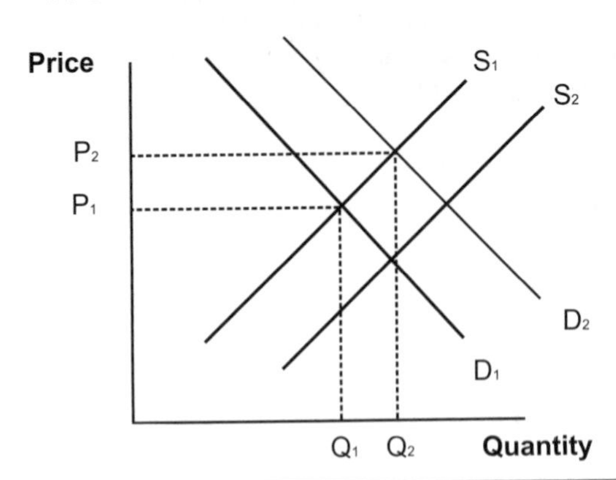

At this equilibrium point the price of brussels will be P_1 and the quantity demanded Q_1. If the demand for brussels increases, (perhaps one of our celebrity chefs comes up with a new recipe) this is represented by a shift of the demand curve to the right to D_2. Now the quantity demanded rises to Q_2 and unless the supply can be increased this will result in a rise in the price to P_2. In the longer term the increase in price may well encourage farmers to plant more brussels. This increases the supply to S_2 and a new equilibrium point where the balance between supply and demand is restored, will be reached.

Supply and elasticity

As illustrated in Figures 5.1 to 5.3, in ideal circumstances (a perfect market) supply and demand will always tend towards equilibrium. So, if demand for crude oil increases its price will rise. The rise in price may well be enough to encourage some producers to increase production and this rise in supply will satisfy the extra demand and the price is likely to fall back to its equilibrium level. However, as we have seen, the property markets are far from being perfect markets and so do not behave quite according to this simple rule. Assume, for example, that following the discovery of North Sea oil, the demand for four bedroom houses in Aberdeen rose. In the case of housing it is not possible to switch on a tap and generate additional supply in the short term. The only ways to increase the supply of housing would involve either building new houses or converting existing buildings from other uses.

Both responses take time. The supply of housing is said to be inelastic; in other words it takes time for changes in supply to respond to changes in demand. Eventually, perhaps higher demand will encourage owners to release land, and developers to obtain planning permission and complete new houses to meet this new demand, but in the interim the increased demand and the short term fixed level of supply will result in an increase in price.

The overall supply of land for property development is fixed. It is said that land cannot be created or destroyed. This is fundamentally true and although it is possible to recover previously unusable land, through the cleaning up of contamination for example, but this has a relatively small impact on the overall supply. Supply could also be effectively increased by increasing the density of development (by building higher for example) but there are generally strict planning controls over how land can be used and these tend to further restrict the supply of land for particular uses in particular locations. This is especially true in the case of the UK with its relatively small land area, existing high density of urban development and consequently strict planning regime protecting the green belt to prevent urban sprawl.

Demand

If anything the demand side of the price equation is even more complex. This is partly because there are so many different factors influencing demand that the valuer needs to take into account. The price of all types of property has risen more or less continuously (apart from the occasional relatively short periods of recession) over the last 150 years or so. This is the result of a more or less continuous increase in demand set against a relatively fixed supply.

This increase in demand has been driven by a number of factors. Most obviously the population of the UK has increased significantly over this period, so there are more people to be housed. At the same time, prosperity has increased as a result of increased economic activity and this has increased the demand for all manner of goods and services, along with increased demand for factories to produce those goods, offices to supply those services, shops to sell the goods and leisure premises where people spend their increased leisure time ... and so on.

This is of course simplistic. An increase in the population may increase latent demand but for this to have any effect on price, that latent demand has to be converted into real demand. It is one thing, say, to aspire towards owning a larger and more expensive house with a large garden, it is another thing to have the means to achieve this. So, a further major constraint on the demand side of the equation is the availability and cost of finance. High interest rates will tend to have a dampening effect on price whereas the willingness of banks and building societies to lend increasing amounts against borrowers' incomes was one of the reasons underpinning house price rises in the late 1990s and early years of the 21st century.

We have seen that over time the increase in population has been responsible for a general increase in demand for most, if not all, sectors of the property markets. However this needs to be explored in more detail. The age distribution of the population, for example, also has an impact. The number of elderly people is increasing and this has led to an increase in demand for housing to suit their particular needs; bungalows, sheltered housing and warden controlled developments for example. Changes in the nature of society are also important. An increase in the number of single people (increasing divorce rates and later marriage or formation of partnerships) has led to an increase in demand for smaller residential units and was one of the factors behind the increase in demand for flats in city centres.

The above examples might all be seen as more global changes impacting on the wider property markets but these demand pushes are not equally distributed and some areas will be more prosperous and attractive than others. There is, for example, greater demand for property in London and South-East England because of relatively high incomes and higher levels of economic activity. This can result in large regional differences in prices for most property types and this is especially true of the residential market for owner-occupation which, as we have seen, tends to operate at a fairly local level. The table below shows regional differences in average house prices in the third quarter of 2008. Average prices should always be treated with caution. Furthermore the data source is HBOS (formerly the Halifax Building Society) and is based on mortgage data and so does not reflect the whole of the residential property market. These average prices, however, serve to illustrate substantial regional variations within the UK.

Region	Average Price
UK Average	£175,143
Yorkshire and Humberside	£128,591
London	£269,723

Regional variations in average house prices, Q3 2008
Source: *http://www.hbosplc.com/economy/latestregionalsummary.asp* accessed 3 December 2008

In the case of residential property, fashion and taste also play an important role in affecting demand. Some locations can become more fashionable than others. The process of 'gentrification', that is the social advancement of an inner-urban area associated with the movement of more affluent individuals into a lower-class area, has seen many once prosperous inner-city locations re-establish themselves as popular residential areas. Islington in London is a good illustration of the process but no doubt every major city will boast similar examples. Similar trends can also be seen in the redevelopment of former industrial areas within major cities. Docklands in London, Salford Quays in Manchester, Albert Dock in Liverpool and the Floating Harbour in the centre of Bristol offer good examples of waterfront development in former derelict dock sites.

Physical changes are also important. One of the most important factors driving demand is the changing nature of the transport infrastructure. The impact of the extension of the rail network during the 19th and 20th centuries is well documented. In modern times, network hubs such as motorway junctions become important locations for distribution warehouses and the accessibility of these locations to large populations has been responsible for the development of a number of large supra-regional shopping centres, such as The Metrocentre in Gateshead, The Trafford Centre in Manchester and Meadowhall near Sheffield.

This overview of the economic forces underlying the demand and supply of property provides no more than an outline and the valuer needs to develop an intimate knowledge of all these factors and most importantly will need to be in a position to assess future trends. Valuers need to be able to undertake in depth analyses of any markets they operate in as well as a wider understanding of prevailing macro economic conditions. Demand and supply in the residential property markets is considered further in Chapter 6 and the commercial property markets are explored in Chapter 7.

Further reading

Modern Economics, Harvey J and Jowsey E, 8th ed, Palgrave Macmillan, Basingstoke, 2007

Urban Land Economics, Harvey J and Jowsey E, 6th ed, Palgrave Macmillan, Basingstoke, 2004

6 Residential Property Markets

House prices

The British seem to have a peculiar obsession with house prices. News headlines report every rise and fall, the organisations that produce statistics on the housing market are guaranteed free publicity with every release of their data and day time television seems to be dominated by cheaply made programmes about the property market.

The table below shows average UK house prices for the years 1950 to 2007 (taken from Table 502 Housing market: house prices from 1930, annual house price inflation, United Kingdom, from 1970 — available on the Department of Communities and Local Government web site *http://www.communities.gov.uk*).

Average house prices 1950 to 2007

	1950s	1960s	1970s	1980s	1990s	2000s
0	£1,940	£2,530	£4,975	£23,596	£59,785	£101,550
1	£2,115	£2,770	£5,632	£24,188	£62,455	£112,835
2	£2,028	£2,950	£7,374	£23,644	£61,336	£128,265
3	£2,006	£3,160	£9,942	£26,471	£62,333	£155,625
4	£1,970	£3,360	£10,990	£29,106	£64,787	£180,248
5	£2,064	£3,660	£11,787	£31,103	£65,644	£190,760
6	£2,280	£3,840	£12,704	£36,276	£70,626	£204,813
7	£2,330	£4,050	£13,650	£40,391	£76,103	£221,580
8	£,2390	£4,344	£15,594	£49,355	£81,774	
9	£2,410	£4,640	£19,925	£54,846	£92,521	

So between 1950 and 2005 the price of an average house went up from £1,940 to £190,760, a factor of £190,760/£1,940 = 98.33 times, an annual increase of 8.7% (Chapter 13 shows how to work this out).

The following chart shows the average house prices plotted against inflation — the data is taken from the Office of National Statistics Retail Prices Index: annual index numbers of retail prices 1948–2007 (*http://www.statistics.gov.uk*).

Figure 6.1 House prices and inflation 1950–2007

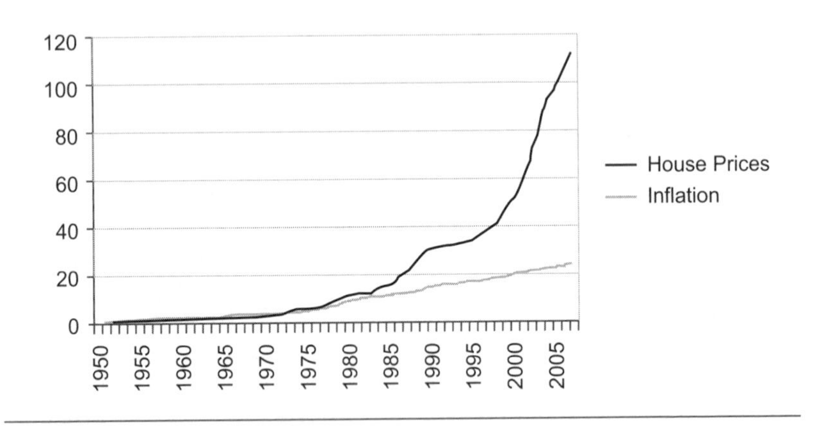

From the early 1970s residential property has been a staggeringly good investment, consistently giving an inflation beating return which, for owner-occupiers, is tax free because no Capital Gains Tax is payable on the sale of a main residence.

Imagine that a couple bought a house in 1997 for owner-occupation at about the average price of £76,000. They had some money available so paid cash. The house was sold 10 years later in 2007 for £221,500 (again about the average price), this resulted in a tax free gain of £145,500 (a return on the investment of 11.3%pa) and they have had a house to live in, in exchange for giving up interest on capital (on which Income Tax would have been paid).

Our owner-occupiers could have done better. Taking out a mortgage of 90% of the price (£68,400) would have required them to pay a deposit of only £7,600. For the sake of simplicity let us assume that the mortgage was interest only, and that the capital would be repaid when the property was sold, leaving a gain of £221,500 – £68,400 = £153,100 from an investment of £7,600. The tax free gain is still £145,500 but the return on the money invested is now 35% a year (this takes no account of the interest paid on the loan, but the purchasers have had free accommodation for 10 years). Using borrowed money (or debt) to multiply potential rewards is called leverage.

Professional movers repeat the leverage trick at frequent intervals as increases in their income allow them to borrow more money, buying more expensive property at each move and thus getting greater gains for their original outlay as well as a better house.

The increase in house prices however has not been steady, the market is cyclic and has been affected by booms and busts. These can be explained using basic economic principles (see Chapter 5).

Supply of housing

The total supply of housing is limited. About half of the 21 million houses in England are over 50 years old.

Year built	All tenures and all types of accommodation
pre 1919	20%
1919–1945	19%
1945–1965	22%
1965–1985	25%
post 1985	15%

Department for Communities and Local Government, Housing in England 2006–07, 2008

Little new land for housing can be created and many towns and cities are protected by green belts which prevent them from expanding, however land can change from one use to another and land can be developed at higher densities. In practice few new houses are built as a percentage of the whole stock, the creation of new stock averaged about 0.7% of the total dwelling stock each year from 1991–92 to 2004–05. The supply of housing at national level is therefore relatively fixed (or inelastic) and rapid changes in the number of properties available cannot be made because houses take time to plan and build.

Long term, government policy can significantly affect supply in four ways: taxation, fiscal policy, planning controls and statute.

Taxation

Taxing development or the sale of development land increases the cost of development and makes building houses less profitable, thus restricting supply. Over the years several attempts have been made by Labour governments to tax the development value of land, starting with the Development Charge in the Town and Country Planning Act of 1947, followed in the 1960s by Betterment Levy, and in 1976 Development Land Tax was introduced. All these measures were abolished within a few years by the next Conservative government. In the meantime landowners avoided the taxes by not selling land, thus restricting the amount of land available for development. In the early years of the 21st century concerns over the shortage of reasonably priced housing for essential workers, particularly in the South East, led to the introduction of affordable housing policies which allowed planners to require developers to provide a proportion of social housing as part of all but the smallest scheme. Making the developer subsidise social housing is effectively a tax, reducing the amount of profit that can be made and discouraging development.

Fiscal or monetary policy

Fiscal policy may make credit hard (or expensive) to obtain. This will discourage new development and make it harder for purchasers to get a mortgage. Alternatively, a relaxed fiscal policy will mean that developers can fund their schemes cheaply and easily, and buyers have no difficulty in raising finance to buy a house.

Planning controls

Planning controls affect supply in a number of ways. Planning permission is required before development can take place. Planners allocate land use so it would not be possible to get permission to build houses in an area zoned for industrial use. Green belts restrict the expansion of towns and cities by preventing development in the countryside around them. Planning policies can also limit changes of use and restrict the intensity of use, by preventing large houses being converted into a number of flats and limiting the number of houses built on a given site by imposing a plot density. In recent years the government has encouraged higher densities. Planning Policy Statement 3 (PPS3) Housing (Department of Communities and Local Government, November 2006) says "Local Planning Authorities may wish to set out a range of densities across the plan area rather than one broad density range although 30 dwellings per hectare (dph) net should be used as a national indicative minimum ...".

Other statutes

Post World War One the Rent Acts limited the rents which could be charged for houses and gave tenants security of tenure, which made it difficult to get vacant possession. The result was that few owners were willing to let houses. Rent control and security of tenure were relaxed in 1988 with the introduction of assured tenancies and assured shorthold tenancies, which allowed landlords to charge the market rent and get possession of their property more easily. This, together with the availability of mortgage finance for residential investment property, led to buy to let houses and apartments becoming a popular investment. About 5% of private sector lettings are still regulated tenancies under the old regime.

Demand for housing

The demand for housing is for occupation and for investment. There are three potential groups of purchasers:

- owner-occupiers, who tend to see their house as both a place to live and as an investment
- property investors, who need demand for occupation by tenants in order to produce an income stream and
- builders, who buy property to renovate for resale at a profit or for letting.

Houses are expensive and purchase usually involves borrowing money. In 2007, only about 11% of owner-occupiers were cash buyers who had purchased their house outright (DCLG, Housing in England 2006–07, 2008, p72). The prevailing borrowing rate and availability of credit therefore have a critical effect on demand.

Government policy, particularly on taxation can affect demand. In the past owner occupiers were allowed Mortgage Interest Relief At Source (MIRAS). This meant that the interest payments on residential mortgages up to £30,000 were reduced to reflect the basic rate of income tax paid on the income, making borrowing cheaper for house buyers. The announcement in March 1999 of the abolition of MIRAS from April 2000 caused a sudden temporary increase in prices as people bought early to escape the change. More recently the possibility of some form of stamp duty relief in the summer of 2008 slowed the housing market as potential purchasers waited to see how much they could save.

Demand for occupation

Everyone needs somewhere to live. The number of households has gone up because of increases in population (the population of the UK in 1950 was just over 50 million, in 2007 it was estimated at just under 61 million and the population is estimated to rise to 65 million by 2016). In addition demographic changes have taken place, including a fall in the number of couples with dependent children and an increase in the number of single person households.

The number of single people, and the size of families will determine the type of housing demanded. The age of the population will affect the type of dwellings for which there is demand because the young will tend to want flats or cheap starter homes, while the old may want bungalows or sheltered accommodation.

Demand for accommodation comes from new households being formed (first time buyers in the owner-occupier market), from households moving up market to larger property in a better area or down market to smaller property in a cheaper area, and from people changing location because of their jobs or for family reasons.

The fact that there is a need for housing does not necessarily translate into demand. The performance of the economy, including the levels of unemployment, wages and interest rates and the availability of mortgage finance, will affect the ability of potential purchasers to actually buy property in the market and the ability of potential tenants to pay the rent.

The market for dwellings for occupation can be split into three categories.

	1953	1981	2007
Owner-occupiers	4.1 million (32%)	9.9 million (57%)	14.7 million (69%)
Private sector tenants	6.5 million (51%)	1.9 million (11%)	2.7 million (13%)
Council/housing association	2.2 million (17%)	5.5 million (32%)	3.8 million (18%)

Department for Communities and Local Government, Housing in England 2006–07, 2008 and S101: Trends in Tenure *http://www.communities.gov.uk*

There have been significant changes in tenure over the last half century, with a rise in owner-occupation and a fall in the number of private sector tenants until the 1980s, when rent control and security of tenure were relaxed. Buy to let mortgages became available in 1996, and the availability of credit coupled with a fashion for buying residential investments (often by individuals seeking to build a portfolio as part of a pension) underlies the growth in the private rented sector since then.

From the 1980s there has been a fall in the number of council and housing association tenants, caused partly because the 'Right to Buy' legislation gave most public sector tenants the right to buy their houses at a discount (by 2006–07 about 1.8 million had done so (DCLG, Housing in England 2006–07, 2008 p73)) and partly because central government restrictions on public spending meant that few new council houses were built, although from the 1980s housing associations have increasingly used private sector funding to provide social housing.

The characteristics of the property

While the characteristics of the property affect demand from both occupiers and investors, it is convenient to mention these factors here, if only because some buy to let landlords have shown little interest in the bricks and mortar they have purchased until they find that the rent collected is less than they had expected, by which time it is rather too late.

The location, the type of the property (terrace/semi-detached/detached, house/bungalow/flat), the construction and design (an avant-garde house may appear to be attractive to its owner, but it might struggle to find a buyer), condition, the accommodation, facilities, parking and so on are all discussed in some detail in Chapter 4.

The age of the property can also have a significant impact on price. Developers will often have a substantial marketing budget and will often offer incentives to purchasers, such as cheap mortgage deals, cars, white goods, cash-back and so on. In addition some buyers much prefer to have a brand new house and will pay a premium to be able to choose the fittings and colour scheme. This can cause a difficulty for agents trying to sell a property on the same estate which is only just second hand, as their client may have difficulty in understanding that the house is worth less than they paid for it a matter of months earlier.

Market changes

Figure 6.2 House price growth in real terms

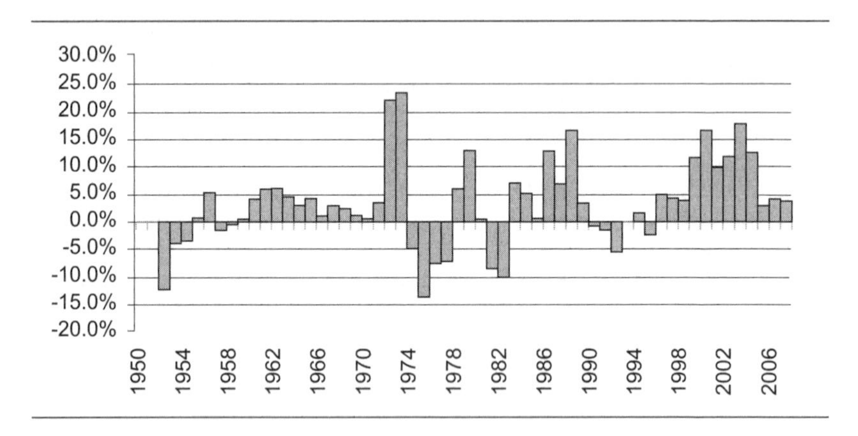

Figure 6.2, which uses the same sources of data as Figure 6.1, shows the cyclic nature of the market.

The property market was relatively stable in the post war period, broadly keeping in line with the rate of inflation, but started to rise in the 1960s as increasing affluence created more demand. Since then there have been four house price booms.

The first occurred between 1970 and 1973. Mortgages were readily available and the interest rate was low, so house prices rose dramatically. In 1973 there was a credit squeeze, interest rates rose and borrowing was limited, resulting in a reduction in the rate of increase in house prices, and in some cases, prices falling.

A second boom happened in the late 1970s. At that time the normal maximum mortgage advance was two and a half times the husband's salary plus half times the wife's salary (a reminder of just how different society was back then), subject to a maximum advance of 90% of the value of the property. Most mortgages were provided by building societies, funds were in short supply and borrowers commonly had to queue for a mortgage.

Poor national economic conditions with higher interest rates and a high inflation rate limited demand at the end of the decade, and by 1980 the housing market was poor.

The market had recovered by the middle of the 1980s, and there was a third house price boom in response to the booming economy at the end of the Thatcher era. In 1986 finance markets were deregulated (the big bang). Building societies were able to demutualise and become mortgage banks. Increasing amounts of money were available to lend, and banks (clearing banks had entered the mortgage market in the early 1980s) and building societies relaxed lending criteria, with mortgages of three times the main income plus one times a couple's secondary income becoming common.

Monetary policy was tightened in the late 1980s as inflationary pressures began to emerge in the wider economy, with mortgage interest rates rising to over 15% in 1990. This helped to push the economy into recession, and the housing market weakened considerably. Prices began to fall as unemployment rose.

Interest rates fell in the middle of the decade, which helped prices recover. The demutualised building societies led the way with loans of 100% of the value of the house or five times a couple's joint income (Northern Rock offered a 125% loan!). Buy to let mortgages became available from 1996, increasing demand from investors. From the late 1990s interest rates fell and there was a fourth house price boom. First time buyers were desperate to get on the housing ladder before prices rose to a level that they could not afford. Lower paid workers, particularly in the South East, were unable to buy houses, resulting in the creation of affordable housing policies. The boom slowed from 2005 but continued until 2007. Buy to let landlords came under pressure as rising interest rates and falling rents made it difficult to service their debts (particularly in some city centres where so many apartments had been built that there was insufficient demand from tenants, so that rental values fell and significant numbers of flats remained unlet).

In the summer of 2007 the credit crunch began. Sub-prime mortgages (loans to house owners who were unable to re-pay the loans) in the United States had been securitised (that is they were converted into financial instruments which could be sold to other banks). The bubble in the US housing market burst and it was suddenly realised that the securitised stock was toxic. Interbank lending reduced significantly world-wide. The Northern Rock, a former building society which had converted to a bank, had expanded rapidly by borrowing on the money markets, the loans could not be renewed and the bank was nationalised. Lending institutions considerably restricted mortgages resulting in the end of the fourth boom. By 2008 mortgage lending had roughly halved, resulting in a very slow housing market, falling prices and developers going bust. The poor market did not improve the problem of the lack of affordable housing as lenders required larger deposits and mortgage interest rates increased. The 100%+ mortgage was no longer available, and high percentage loans were available only at higher interest rates. As the UK economy headed into recession in 2008 the housing market was already in trouble because of the credit crunch. Experts who had been consistently over optimistic continued to downgrade their views of the market. By the autumn of 2008 none of the demutualised building societies that had become mortgage banks were still in business — they had all been taken over or nationalised.

While the housing market has been characterised as boom and bust, prices rarely fall in money terms. Most owners don't sell in a poor market if they have a choice, so the market can be dominated by a limited number of transactions which will include a higher than normal proportion of properties which have been repossessed by lenders when the borrower has failed to pay the mortgage (repossessions).

Local markets

Having painted a picture of the housing market at national level, it is important to understand that the housing market is broken up into a range of sub-markets which, against the backdrop of the national and global economy, may not move in the same direction. For example:

- London and the South East normally have greater demand than the rest of the country
- migration can cause a significant increase in demand. A major firm moving into an area will bring with it higher paid staff who will often dominate the top end of the property market causing significant increases in prices. The effect will be magnified if the firm is moving from a high value area to a lower value area, as the

staff being relocated will normally want to buy houses at similar prices to the ones they are selling
- the market in some rural and seaside areas is distorted by demand from people seeking second homes
- although new build represents only a small percentage of the total stock at national level, locally it can flood the market, as happened in some Northern cities where large numbers of apartments were built in the early years of the 21st century. The result was high levels of vacant flats (voids), falling rental values and buy to let landlords who could not service their loans
- the state of the economy locally may be different from the national economy. In the late 1980s, at a time when most of the housing market was booming, coal mining areas had been badly affected by the closure of many pits, as a consequence their housing markets were very depressed.

Demand for residential valuations

In 2004–05 about 6% of owner-occupiers had moved during the previous year. The median time owner-occupiers had lived in their properties was 11.5 years. (Department for Communities and Local Government, Housing in England 2006–07, 2008). Given that there are 14.7 million owner-occupied dwellings in England (with a value of over £3,000 billion) there is normally a considerable demand for valuations of residential property in connection with market transactions.

Valuations for sale are often provided as informal guidance offered by estate agents as part of their advice on marketing. However, formal valuations may be required in connection with a proposed sale by trustees, mortgagees in possession (where the borrower has defaulted and the lender intends to sell the security to recover the outstanding debt and expenses) and some public sector bodies.

In normal times only about 10% of buyers pay cash, and mortgage lenders will require a valuation to ensure that the property offers satisfactory security for the loan. The *RICS Valuation Standards*, 6th Edition, UK appendix 3.2 provides the RICS mortgage valuation specification for this work.

For many buyers their house is the most expensive purchase they will ever make. Some lenders provide the borrower with a copy of their valuation report. Many buyers assume that if a bank or building society will lend money on mortgage the deal is "OK", and the cases of *Yianni* v *Edwin Evans* [1981] 2 EGLR 118 and *Smith* v *Eric S Bush (a firm)* [1989] 1EGLR 169 show that a valuer owes a duty of care to the borrower and is potentially liable to both the lender and the borrower for negligence. This applies regardless of whether or not the borrower has been given a copy of the valuation report.

About 15% of buyers have a Homebuyer Survey and Valuation (HSV) carried out, often commissioned via their lender. The *RICS Valuation Standards*, 6th ed, UKPS 4.1 requires valuers accepting instructions to provide this service to comply with the practice notes published by the RICS. The HSV is to be replaced in 2009 by the new Home Buyer Report (HBR). Only about 2% of residential purchasers commission a building survey, which may include valuation advice.

Housing associations sometimes sell houses and flats on a shared ownership basis. The purchaser buys a share of the property (normally 25% or 50%) and pays rent on the remainder. The owner/tenant usually has the option to buy additional shares (commonly in multiples of 25%) until they own 100% and become an owner-occupier. A valuation is required each time a share of the property is sold, and the *RICS Valuation Standards*, 6th ed, UKGN 2 and UKPS 4.3 should be consulted when carrying out this work.

Some owner-occupiers will ask their lender for a further advance, that is an additional loan, on their mortgage. This might be so that they can carry out repairs or improvements, although some lenders will allow a borrower to release some equity and not ask questions about what the money will be used for. The lender will require a new valuation unless they are confident that the original report will cover the total loan outstanding. Some lenders now use automated valuation models when considering applications for a further advance and do not issue instructions for a formal valuation.

Registered social landlords (housing associations) may use their housing stock as security for loans. The *RICS Valuation Standards*, 6th ed, UK appendix 3.3 gives guidance on preparing valuations for this purpose.

Owners of houses held on long leases usually have the right to buy the freehold of their property (enfranchise), and owners of flats held on long lease often have the right to extend their leases and to collectively enfranchise (that is to buy the freehold with other leaseholders in the same block). *Valuation Principles into Practice*, Hayward R (ed), EG Books, 6th ed, 2008 has a chapter on this subject.

Asset valuations, that is valuations to be included in accounts, will be required for residential property owned by companies, housing associations and local authorities. Special rules apply to the valuation of property used for social housing (details can be found in *RICS Valuation Standards*, 6th ed, UKPS 1.12 Local authority asset valuations, UK appendix 1.5 Valuations of local authority assets and UKPS 1.13 Valuations for registered social landlords).

Various statutory valuations may be required for residential property, including for Council Tax, Inheritance Tax, Capital Gains Tax, compulsory purchase and where council tenants exercise their right to buy their houses or flats at a discount. The Valuation Office Agency manuals on these valuations are available on their web site *http://www.voa.gov.uk* under the publications tab.

Further reading

RICS Valuation Standards, 6th ed, 2008
Valuation and Sale of Residential Property, Mackmin D, 3rd ed, EG Books, 2007, particularly Chapters 1 and 2.
Valuation Principles into Practice, Hayward R (Ed), 6th ed, EG Books, 2008 particularly Chapters 2 and 3.

7 Commercial Property Markets

This chapter examines the size and main categories of the UK's commercial property markets. It outlines some of the key factors affecting the level of market rents, including landlord's outgoings and rent review adjustments, and factors affecting the level of market values (that is capital values).

Together with earlier chapters, it sets the context for the chapters that follow which examine the principles and rationale of the five methods of property valuation. The comparative method is normally chosen to assess the market rent of commercial property and the investment method is normally used to assess its market value. The investment method presents a range of relatively challenging conceptual and methodological issues and these are examined in the final six chapters of the book.

The media often refer to 'the property market' as if it were a single entity. Such a phrase is nonsense, because, as we saw in Chapter 5, there are a multitude of markets in property. This is certainly true of commercial property, which includes shops, offices, factories and warehouses.

Across the commercial property sectors, as with the residential property sector (see Chapter 6), there are two major categories. These are occupation markets and investment markets. Commercial property markets may also be distinguished by reference to property type, size and location.

Occupation markets

If a company has a long term commitment to operate from particular shop, office or industrial premises, it will probably wish to buy a freehold or long-leasehold interest (for example 125 years). This will require the commitment of capital, probably by way of mortgage debt, that is needed to facilitate such a purchase. Such companies are in the *owner-occupation* market which overlaps with the *occupation* market and is linked to investment markets (outlined below).

Alternatively, a company may decide that it can invest its resources more effectively by acquiring plant and equipment for the business, rather then by buying premises. In these circumstances, the company will decide to rent premises rather than acquire a freehold or long leasehold interest. Renting is generally regarded as being a more flexible way to gain occupation of business premises and, although there will be the potential disadvantage of a possible rising rent at each rent review, the decision to rent frees a company from the worries of outright ownership and allows it to concentrate its resources on its business activities. Some companies which own the freehold of the premises they occupy may sell their interest to an investor and then take a lease of the property, a practice known as sale and leaseback, to allow them to obtain capital to invest in the business rather than in property.

Investment markets

In response to the demand for rented business premises, there are several types of organisation that own freehold and long-leasehold interests in commercial property in order to grant occupation leases.

The Investment Property Forum's report, *Understanding Commercial Property Investment: A Guide for Financial Advisers*, 2007 ed, estimates that, as at the end of 2005, the main investors in UK commercial property were:

Type of investor	Approximate share
UK financial institutions, such as pension funds and insurance companies	28%
Overseas investors	15%
UK private property companies	15%
UK listed property companies	14%

Limited partnerships	7%
Traditional estates and charities	5%
Property unit trusts	4%
Unitised and pooled funds	4%
UK private investors	3%
Other investors	3%

The report also estimated the size of the main investment market sectors, as at the end of 2005, as:

Sector	Value
Commercial property	£762 billion
Private residential market	£3,400 billion
Equities quoted on the London Stock Exchange	£1,781 billion

The Office for National Statistics figures show that more than 60% of Britain's wealth was tied up in property at the end of 2007. However, "2008 witnessed a sea-change in the global investment environment, following what the International Monetary Fund has labelled 'the largest financial shock since the Great Depression'" (*Money into Property*, DTZ, 2008). The RICS Commercial Property Forecast: December 2008 reported commercial property prices were over 25% down on the June 2007 peak and that further declines were expected. So, although the property market crash seriously reduced its value, property remains a key element of the wealth of the UK, particularly bearing in mind that the value of the other investment sectors fell at the same time.

The figures show that distinct markets for shops, offices, warehouses and factories (see below) account for about 80% of the commercial property market and around half of this is investment property, where it is rented to tenants by landlords. The rest is owner-occupied, mainly by private companies and public sector or non-profit making organisations.

Other market distinctions

Property markets can also be distinguished with reference to the following characteristics.

Property type

No two properties are likely to form part of the same market unless they are of the same type. Local authority planning policies may ensure that properties which appear similar to each other are, in fact, in different markets. For example, a shop in a high street may have its value increased if planning permission is granted for its use as a fast-food restaurant and a warehouse may be enhanced in value if planning permission is granted for retail use. It should be emphasised, however, that the mere granting of planning permission does not itself add value to a property, or transfer it into a different category of market. There has to be demand for the alternative use.

Property size

A department store can hardly be in the same market as a small corner shop. So, size plays a part in distinguishing property markets from each other. However size is a continuum, and it is difficult to always say with certainty where the boundaries lie between markets for properties of different size. Size is also obviously a major factor influencing value but it should be noted that very large properties may have limited markets. For example, a £60m office investment will only be of interest to a limited number of investors, and a very large industrial unit would only attract a small number of occupiers.

Property location

Properties used for similar purposes and of a similar size may be in different markets if their locations are sufficiently distinct. It is sometimes difficult to identify the geographical extent of some commercial markets. Depending on the type and size of property being considered a market may be local, regional, national or even international in scale.

For example, the market for small office suites may be generated purely by local factors. Small firms of solicitors, accountants and other professional practices may generate a demand for such space from within a town or city, with virtually no demand being generated for such space from outside the area. However, newer office suites, in purpose built modern buildings, will probably appeal to a wider market. Large companies, for example insurance or call centre companies, nationally known firms of accountants, or government departments, may be in the market for such premises in order to establish regional offices. Such organisations may therefore compare the available property in two or more centres within a region before deciding where to locate and, in that sense, commercial property of a certain type and size in adjacent towns or cities could be said to be in the same economic market at a regional level. Similarly, regional markets in properties suitable for distribution centres will tend to be located along motorways, with distribution companies searching for premises near motorway intersections.

Markets can also exist at a national level. For example, Nissan located its UK car manufacturing plant at Washington in North East England after scouring virtually the whole of the UK for a suitable site. In a global economy, organisations may seek to locate their headquarters in Europe and compare several possible locations in major European cites. In that sense the market for major buildings in London may be in competition with the same market in Paris, Berlin or Zurich and other cities offering similar facilities.

Firms of valuers are frequently organised so that the individual staff members can gain experience and specialise in the different markets outlined above. If a firm is sufficiently large it is likely that one or more valuers will specialise in shop rental markets, office rental markets, investment markets, and so on.

The level of commercial property values

Information about the major property markets can be obtained from the research publications of the major firms of commercial real estate consultants, for example, Jones Lang LaSalle, DTZ, Cushman & Wakefield LLP and CB Richard Ellis, and government agencies, such as the Valuation Office Agency (VOA). The journals and electronic data providers which serve the property market also publish useful information regarding the commercial property market. Major information sources include the Estates Gazette and Estates Gazette interactive (EGi) (*www.egi.co.uk*) together with Focus Commercial Real Estate Information (*www.focusnet.co.uk*).

Examples of prime market rents

The VOA, an executive agency of HM Revenue and Customs, provides useful reports, twice a year, drawn from local evidence of market rents, including all major sectors of the property market across the UK. The table and notes below are extracts of this data for prime property, illustrating typical ranges of market rents for four sample cities in England: Bristol, London, Reading and Sheffield, as at July 2008 (see *www.voa.gov.uk/publications/index.htm* for current data).

Location	Shops[1] £/m²pa	Offices[2] £/m²pa	Industrial[3] £/m²pa
Bristol	2,200	270	90
London (City)	3,500	645	–
London (Southwark)	–	–	130
Reading	2,350	225	130
Sheffield	2,500	165	65

Notes

1. Shop figures are Zone A rental values (see Chapter 9 for an explanation of zoning), on full repairing terms, reported by VOA District Valuers (DVs) for prime positions in principal shopping centres. Rents have been analysed on a zone pattern of 6.10m zone A, 6.10m zone B. This slightly odd zone depth is a throwback to the old imperial measure of zone depth based on 20 ft.
2. Office figures are headline rental values, on full repairing terms, with no inducements reported by DVs for town centre locations. They assume a self contained suite, over 1,000m² in an office block erected in the last 10 years, with a good standard of finish, a lift, good quality fittings to common parts and limited car parking available.
3. Industrial figures assume an industrial estate location, a letting on full repairing and insuring terms, modern construction but not of high-tech design, heated by free standing heaters, small starter units of 25m² to 75m², steel framed, concrete block or brick construction, built in terrace layout and let on weekly terms.

Property rents generally reflect the profitability of the businesses that occupy various types of premises (see Chapter 10). Changes in rents reflect changes in the health of the economy because such changes influence supply and demand. So, for example, following the October 2008 global financial crisis and the potential depth of the UK economic downturn and recession, Jones Lang LaSalle (a global real estate services firm specializing in commercial property management, leasing, and investment management) forecast that City of London prime office rents would fall by close to a third by 2010 from their peak at the end of 2007. In other words, rental falls in the City could be similar to those that followed the dot.com collapse when rents in the City fell from around £737/m^2 pa in the third quarter of 2001, when the new internet-based companies bubble burst, to £516/m^2 pa in the third quarter of 2005.

The demand for rental valuations

The main demand for rental valuations, from both landlords and tenants, arises when there is:

- a first letting of a new or refurbished building
- a re-letting of a building vacated by a previous tenant
- a renewal of an expired lease when a tenant wishes to remain in occupation of premises
- a rent review due because of a provision in the lease.

In the first case a valuer will probably have been engaged from an early stage in planning the development to advise on the likely value of the project. The valuer will be retained by the developer client for the duration of the development to secure lettings. A tenant may seek the advice of a valuer to negotiate a rent payable to ensure it is not more than the market rent.

When letting space in a new development, or when re-letting space, the landlord may agree to allow the tenant a rent-free period of a few weeks, or months, to fit out the premises to a satisfactory standard. The length of the rent-free period will vary and be longer if the property has not been easy to let at the required rents, with rent free periods of up to three years being reported in the winter of 2008–09. Long rent free periods expose the landlord to a risk that the tenant may default before any rent is paid.

Sometimes, in the case of retail developments, the rent which is to be charged upon first letting will not be based on the concept of market rent. Instead the rent will be based on a formula related to the turnover of the tenant. Turnover is the total amount of sales the tenant achieves from the premises and such rents are referred to as percentage rents or turnover rents and are frequent in large regional shopping centres as they allow the landlord to participate directly in the success of the centre, without having to wait for rent reviews every five years. In order to protect the landlord from possible poor trading practices by a tenant, there will normally be a base rent which represents a minimum rent payable. The valuer would need experience of retail markets of this sort before offering reliable advice on the percentage figures to be applied.

When re-letting space in a building which has already had previous occupiers, the valuer will normally value to market rent, unless the landlord has a policy of letting on an alternative basis, such as a percentage rent. Again the landlord may have a policy of allowing a rent-free period, especially if the market at the time of the letting generally favours tenants, as it will allow the landlord to maintain a higher 'headline rent' (see Chapter 2, Glossary).

The valuation of a commercial property upon the renewal of a lease after the leasehold interest expires, with the tenant retaining occupation, is more complex. Statutory intervention, in the form of the Landlord and Tenant Acts 1927 and 1954 (as amended), means that although the valuation may be to market rent the valuer has to take account of the existing relationship between the parties in a way determined by statutory provisions. If a dispute arises regarding the level of rent, or other terms of the new lease, valuers representing the landlord and the tenant may find themselves giving evidence in court proceedings and the quality of their valuation evidence will be closely scrutinised.

Lease flexibility and break clauses

In the UK the dominance of institutional landlords, from the late 1960s to the early 1990s, caused most occupational leases to be relatively long term commitments, with lease terms of from 15 to 25 years, with five year rent reviews. The UK property market has been slow to offer flexible lease terms but tenants have increasingly pressed for more business friendly terms. The government has also expressed concern that UK businesses, seeking to implement efficient property strategies, were not operating on a level playing field when compared to their international competitors, who benefit from shorter lease terms. Collaboration between commercial property professionals and industry bodies representing both owners (landlords) and occupiers (tenants) has resulted in a voluntary guide: The Code for Leasing Business Premises in England and Wales 2007.

It is now common to find leases incorporating a break clause which allows the tenant, and sometimes the landlord, to end the lease before its expiry date. Such a clause may affect the market value of a property because the market norm is to assume that the tenant's break option will be exercised and this will impact upon the landlord's continuity of income.

For generations, rent on commercial property has been paid on a quarterly in advance basis. In recent years the British Retail Consortium has pressed for commercial landlords to accept rent on a monthly in advance basis, to help tenants better manage their cash flows. The 2008 credit crunch added momentum to the campaign and, as a result, landlords are increasingly accepting this smoothed form of rental payment. Some landlords believe that monthly payment of rent offers them early warning of a tenant's financial difficulty and an opportunity to offer support, a better option than facing a bankrupt tenant and a vacant building.

The influence of lease clauses on rental value

Chapter 1 considered the nature and form of a lease and the typical covenants in a lease. Chapter 2 set out the definition of Market Rent which indicated that the lease terms have an effect on market rent; the more responsibility placed on a tenant the more the market rent will be reduced.

There are two possible reasons explaining why a landlord would insert a clause that lowers the rent that can be charged for the property.

- The landlord may want to control how the property is used and may only be able to achieve this by accepting a lower rent. For example, the landlord of a shopping centre may wish to achieve a specific mix of shops in order to attract the largest possible number of shoppers. To do this the landlord may have to let some shop units to tenants whose business will not allow them to pay the highest possible rent. In such cases, the landlord would include a user clause, which would allow the tenant to use the premises only for a specific retail trade.
- The landlord will happily settle for some reduction in rent if, as a result, expenditure commitments on maintenance, insurance or other outgoings are avoided; or if the tenant covenants to carry out improvements.

In the majority of lettings the landlord will hope to negotiate a *full repairing and insuring (FRI) lease*, that is a lease that places responsibility for all the repairs and insurance of the premises on the tenant. If the building is let to a single tenant it is normally easy to negotiate FRI terms in the lease. If, however, a building is let to several different tenants with each occupying part of the building, it would not be practical to ask each tenant to be responsible for repairing and insuring their part of the structure, in addition to the internal maintenance of the space occupied. In this case a service charge clause would be inserted in the lease which requires the tenant to reimburse the landlord for the cost of repairs and insurance for the whole building and for expenditure on maintaining, heating and lighting common areas, for example lifts, stair wells, toilets and corridors, and providing reception and security services. Typically, the reimbursement is borne by the tenants in proportion to the amount of space occupied.

All covenants are open to negotiation and tenants are not obliged to accept excessively burdensome lease terms. They always have the option of renting premises elsewhere; a market with weak demand will give tenants a strong negotiating position. In the 1980s, better quality commercial properties were normally let on a standard institutional lease for a term of 20 or 25 years. The length of new leases has been shortening since the mid-1990s as tenants are increasingly reluctant to enter into long term financial commitments. Crosby and Murdock in 'Monitoring the 2002 Code of Practice for Commercial Leases', March 2005, reported that in 2003 the average lease length for retail property was 9.5 years, for offices 6.7 years and for industrial property 6 years.

Outgoings and rent

Outgoings are costs incurred by the owner of an interest in property, usually calculated on an average yearly basis for such things as property repairs, insurance, management and, if appropriate, rates payable to a rating authority and rent payable to a superior legal interest holder, as specified in the lease.

Wherever practical most commercial properties will be let on FRI leases, with the landlords enjoying net rents, apart from management costs. Occasionally, however, the lease will specify that the landlord is responsible for some, or all, of the repairs and insurance. In this case it will be necessary to assess the average annual cost of each of the landlord's outgoings and to deduct them from the gross rent to assess the landlord's net rental income before the income is capitalised. Also, in order to advise tenant clients of their total annual cost in use of occupation, the valuer may need to assess the level of annual outgoings and service costs before adding them to annual rent payable.

The main outgoings the valuer may be expected to assess when undertaking an investment valuation are repairs, insurance and management.

1. **Repairs**

   While it is difficult to assess the annual cost of repairs, an experienced valuer may be able to judge the cost of the landlord's liability for repairs. If necessary, a cost in use assessment will underpin the figure. This sets out a planned maintenance programme, scheduling all recurring annual works and all works needing less frequent action. Future intermittent recurring costs are expressed as an annual cost and added to regular costs to give a total average annual sum.

   Valuers engaged in desk top valuations may need to take a short cut to arrive at a notional figure for landlord's repairs. One such rule of thumb is to take a percentage of market rent. Such an approach should be used with extreme caution because there is no relationship between repair costs and rental value. The difference in the maintenance cost of a small £100,000pa airport retail kiosk and a large £100,000pa airport warehouse is enormous. The unwary valuer will deduct 5% of rent from both properties as an allowance for repairs. This is clearly a nonsense!

2. **Insurance**

   Under the terms of a lease, responsibility for insuring a building for losses in the event of fire or due to a natural disaster, may rest with either the tenant or the landlord. With FRI leases, or where the landlord is allowed to recover the insurance premium from the tenant, there will not be any deduction from rent before capitalisation. With lettings on the basis of the landlord having responsibility for insurance, the valuer will need to estimate the annual insurance premium and include it in any deduction from gross rent to produce a net income before capitalisation. The amount will normally be based on the premiums payable on a cost of reinstatement figure (see the final section of Chapter 12). Typical insurance premiums might range from £1.50 to £3 per £1,000 of rebuilding cost, with the figure being affected by the risks associated with the type of use of the building. The cost of insuring for £1m might be £1m × say £2.50 per £1,000 = £2,500pa.

Valuers sometimes take a percentage of the market rent as a short cut rule of thumb. So, insurance at 2% of a market rent of £100,000pa = £2,000pa. Such an approach is technically incorrect as market rent and insurance reinstatement cost are not related.

In addition to building insurance, the tenant will also be expected to take out insurance cover for occupier's liability and for potential loss of the building's contents.

3. **Management**

Property management includes collecting rents, carrying out rent reviews, letting and re-letting properties, renewing leases and arranging insurance. Its cost can range from less than 0.5% to 15% of rent collected depending on the extent of management work envisaged. Lower rates apply to single tenant properties let on FRI leases, while higher rates apply to property let on internal repairing (IR) terms, with the landlord responsible for external repairs and insurance.

Some valuers suggest that with FRI leases the cost of management is nominal and is taken care of in the all risks yield capitalisation rate. In other words, using the maxim as you devalue, so shall you value, if the valuer analyses market sales ('devalues') without deducting management from FRI rents to assess capitalisation rates, the valuer should not make an adjustment for management when using the resulting capitalisation rates to find the market value of similar property with FRI leases.

Other valuers, however, now make an explicit deduction for management for all investment valuation work, rather than letting the all risks yield capitalisation rate deal with this variable.

4. **Other possible outgoings**

Rates

Non-domestic rates (often called business rates) are levied on all occupiers of non-domestic property and are their contribution towards the costs of local authority services. Every five years, the VOA has a statutory duty to carry out a revaluation of all non-domestic and business property in England and Wales which ensures that they reflect changes in the property market. Rateable Value (RV) is a key factor in the calculation of business rates but it is not what occupiers actually pay. In broad terms the rateable value is a professional view of the annual rent for a property if it was available for letting on the open market on a set date. All current rateable values are based on a valuation date of 1 April 2003. The next revaluation is due to come into effect on 1 April 2010 and all properties will have their rateable value based on market rental values at 1 April 2008. This will remain effective for five years, unless changes to the property which affect the rateable value require a new assessment to be made. The tax is levied at a percentage figure of the rateable value of the property and is expressed as pence in the pound or rate in the pound. It is set each year by central government and varies slightly from year to year. For example in England the standard multiplier for 2008–09 was 46.2p. So, for a property with a rateable value of £10,000 the local authority will multiply the RV by 46.2p to get a rates bill for the year of £4,620, excluding any discounts or reductions that may be applicable.

Tenants normally occupy on exclusive terms (the rent excludes any payment of rates). This means that in normal circumstances the investment valuer can ignore the cost of rates when valuing a landlord's interest, because the tenant occupiers will pay rates directly to the local rating authority. Exceptionally, a landlord may agree to pay rates on behalf of a tenant and the rent will be described as 'inclusive' of rates. In agreeing a rent on inclusive terms an excess rates clause in the lease may allow for future rates increases to be recovered from the tenant. Deductions may also be made if the rent is inclusive of water and sewerage rates.

Voids

When valuing property let to multiple tenants, for example a large multi-tenant office block or a large covered shopping centre, it is unlikely that the property will ever be fully let. In these cases a voids allowance might be made by capitalising only a proportion of the market rent. With a vacant single tenant property, if there is a strong possibility that it will take six months before a tenant would start to pay rent, it might be valued by deferring the capitalisation of income for six months (see Chapter 14 regarding deferring using the Present Value of £1).

Usually valuers make no explicit allowance for voids in these circumstances on the grounds that their analysis of comparable investment sales data does not make an explicit adjustment for voids and, under the principle of as you devalue, so shall you value, the all risks yield used to capitalise rental income allows for the risk of voids.

It is certainly important not to under-value by adopting both approaches and double counting the allowance for voids!

Outgoings and residential investment property

In the past it was usual to deduct outgoings from the rent when valuing residential investment property; it is now common to capitalise the gross rent received, allowing for outgoings in the all risks yield.

Valuations for rent reviews

The purpose of a rent review is to adjust the rent to the current market level at the review date. The rent will only be reviewed if the lease explicitly provides for a rent review to take place. In general, a rent on review is decided by negotiation between the landlord and the tenant and by reference to the terms of the rent review clause in the lease. This in turn will be based on evidence of comparable market rents. UK commercial leases traditionally provide for upwards-only rent reviews. In this case, the reviewed rent will not be less than the rent currently being paid, even if the market rent has fallen below the current rent payable. Here, the parties will agree to continue the same rent that the tenant currently pays. Open rent review clauses, which allow the rent to reflect both a rise and a fall in the market, are becoming more common as market conditions are putting tenants in a stronger negotiating position. Less commonly, leases may specify the amount of rent payable from each review date or link them to an index, such as the retail prices index (RPI). The problem with this approach is that indexes such as the RPI do not necessarily reflect changes in property market values and so indexed rents can, over time, become quite distorted. Annual increases based on RPI are sometimes agreed between five year reviews to market rent.

Rent reviews are a vital part of the performance of commercial property investment market and the level of rent in the occupation markets. A rent review valuation is a hypothetical valuation. It is not a real market situation. In the real market the prospective parties to a lease are usually free to walk away and a deal will only be struck when a willing tenant and willing landlord agree. A rent review clause attempts to imitate this scenario, but the parties are tied by the terms of the lease and their respective valuations will never be tested by the market. If they cannot agree a figure for the rent their dispute will usually be decided by an independent third party.

Most commercial leases exceeding five years in length will include rent reviews at five year intervals, although three year reviews may be found, particularly for secondary investments. The basis for assessing the rent on review, and the procedure of the review, will be set out in the lease. The rent will usually be assessed on the assumption that the property is vacant and available to let, at the rent review date, in the open market, on the same terms as the existing lease. In order to reflect the position of a hypothetical tenant in the market for the property, three elements must be ignored, as they might otherwise have an effect on rental value. First, the actual occupation of the tenant is disregarded, if not it might be argued that the tenant's bid would exceed market rent in order to remain in occupation. Second, for similar reasons, the goodwill generated by the tenant's business is disregarded. Finally, it is usual for the rent review clause to specify that any improvements lawfully carried out at the tenant's expense during the term of the lease, other than as an obligation to the landlord, will be ignored in so far as they might add to, or reduce value. The reason for this is to prevent the tenant from paying twice for the improvements, once through the original capital expenditure and then again through the rent.

Investment market values

The investment market for commercial and industrial property is made up largely of financial institutions. Their view of the likely investment performance of property investments will determine the values they place on those investments. They normally try to invest in *prime* property, defined as the highest quality property of its type in the best locations. Secondary property is investment property in a less favourable location than prime property.

The profile of investors changes over time and investment in real estate has become a global activity. For example, CB Richard Ellis's Global *ViewPoint Report, Autumn 2008* analysed how Sovereign Wealth Funds (SWF) might impact the commercial real estate market in the longer term. SWF funds derive from a country's money reserves, which are set aside for investment purposes to benefit the country's economy and citizens. For example, United Arab Emirates (UAE) relies on its oil exports for its wealth, so it invests a portion of its SWF fund reserves in other types of assets that can act as a shield against oil-related risk. The amount of money in these SWF funds is substantial. As at May 2007, for example, the UAE's fund was worth more than $875 billion.

The objectives of the holders of investment property are to benefit from both the rental income which is paid by the occupier, and any capital growth which they anticipate from their ownership of the property. As we shall see in Chapter 18 the investor will compare the returns available from property with those available from other investments and, generally speaking, if greater returns are anticipated from holding a property investment than are available elsewhere the investor will acquire the property investment.

Investment returns are generally measured in terms of yields. For example, an organisation which acquires a property investment for £10m will obtain a yield of 5% if the property is let at a rent of £500,000pa. However, the overall yield during the first year of ownership may be considerably greater than 5% if the market value grows during the year. For example, if the market value of the property investment has grown from £10m to £11m after one year of ownership, the investor will have enjoyed a 10% growth in the market value. This will make the overall return during the year 15%, and if no such return could have been obtained from any other from of investment the investor would be well satisfied with this return.

Most investors in commercial and industrial property tend to think in terms of holding their property investments for several years. Buying and selling property investments can be a slow and expensive process, discouraging speculative trading. The organisations that actively acquire property investments generally have long term commitments, to pensioners or insurance policy holders, which encourage the taking of a long term view of their investments. As long as property can be expected to show increased capital values and rental returns in the long term it is likely to retain its attraction to investors of this sort. The Investment Property Forum's report, *Understanding Commercial Property: a Guide for Financial Advisers, 2007 Edition*, estimates that over 35 years commercial property produced annualised returns of 12.2%, ahead of both gilts (government stock) and cash deposits.

Property yields do not stay the same over time. They rise and fall with the property market cycle, as investors gain or lose confidence in the relative risks associated with property investment and its prospect for a rise or fall in value. Consider the following graph of changes in average all-sector prime UK property yields.

Figure 7.1 Yield and interest rate movements

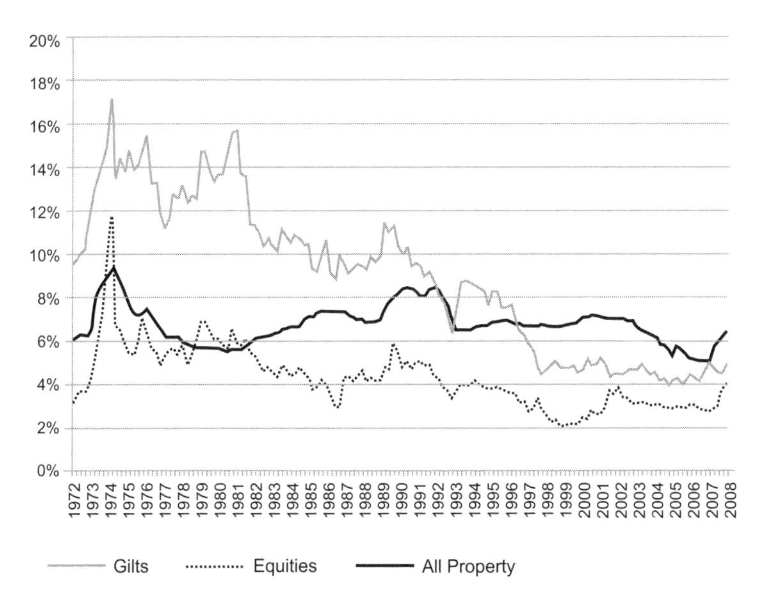

Source: Cushman & Wakefield, Marketbeat UK, June 2008

- From 2003 until 2006, the nominal income return on all asset classes was on a downward trend reflecting growing market values. After 2007 income returns increased with falling market values linked to the change in market cycle which was reinforced by the 2007–08 credit crunch and seizing-up of global money markets.
- From 1980 to 1986 and 1989 to 1992 UK property values were falling.
- A reverse yield gap between property yields and the yield on gilts (fixed interest government stock) from the 1960s to 1997 reflects investors' beliefs that generally property rental incomes would grow.
- When the reverse yield gap was large, for example in 1980, the anticipated growth in rental income was relatively high.
- When inflation is relatively high the yield on *fixed* income government bonds (gilts) tends to be high.
- When the yield on property was relatively high in 1974–75 and 1991–92, the market value of property was relatively low.

- UK property yields since 1997 are generally higher than equities or gilts. This reflects the relatively poor liquidity of property investments and perhaps weak rental value growth potential. In a low-inflation environment, this high yield is one of commercial property's attractions.
- The real estate market peaked in 2007 and its cyclical downturn was greatly reinforced by a sharp tightening in the availability and cost of credit in the money markets and the widespread deterioration in market sentiment initiated by the sub-prime mortgage crisis in the United States.

Investment market yields

High demand for property tends to result in low investment market yields, while property least in demand will tend to have high investment market yields.

Here is a snapshot of average prime yields reported in Cushman Wakefield Marketbeat, Q3 2008 for June 2008:

Shops	5.63%
Industrial	7.08%
Offices (all UK)	6.66%
Offices — Central London	6.05%

The yield for secondary quality properties will be substantially more that the percentage figures above, possibly ranging from 7% to 20% depending on type of property, quality of location and the tenant's covenant.

Cost of buying and selling

The cost of buying and selling property is much greater than for other assets. Professional fees on property transactions are normally estimated at about 1.8% of the price to cover legal costs and agent's fees. In addition, UK buyers pay Stamp Duty Land Tax: currently 4% on transactions of over £500,000, 3% for £250,000 to £500,000 and 1% for £120,000 to £250,000. These costs are normally all reflected in the all risks yield, although some valuers adjust comparables and valuation figures to reflect the costs of acquisition.

Social and economic changes

Valuers must understand the effect of the changes in society and the economic conditions that affect the supply of and demand for property for both occupation and investment. New lifestyles, working patterns and technologies, together with the rise of sustainability concerns are changing the design and demand for commercial buildings. New financial instruments are making property investment available to a wider group of investors and rapid changes in the global economy are having dramatic effects on the level of property market rental and capital values. So, commercial property valuation advice remains a key business service for investors, business occupiers, public sector organisations and private individuals who are increasingly conscious of the importance of real estate to the bottom line of their business.

Further reading

Code for Leasing Business Premises in England and Wales 2007 (see *http://www.leasingbusinesspremises.co.uk*)

Marketbeat: An Overview of the UK Property Market, Cushman Wakefield Q3 2008.

Understanding Commercial Property Investment, a Guide for Financial Advisers, Investment Property Forum, 2007

Valuation Principles into Practice, Hayward R (ed) EG Books, 6th ed, 2008, particularly Chapters 4 to 7.

8 Introduction to Methods of Valuation

This chapter aims to give a brief overview and flavour of the five methods used to value property and outlines how valuation calculations are presented. It explores which methods are used to value different types of property and how methods may be used together to help formulate an opinion of a property's value. Later chapters will examine each of the methods in more detail.

The five methods are the *comparison, profits, residual, contractor's* and *investment methods.*

Sometimes other names may be used, such as *accounts, discounted cash flow, depreciated replacement cost* and *real value* approaches, but they are all variants of one of the above methods.

A valuer may use one, or more, of the five methods of valuation to estimate either the market value (a capital value) or the market rent (a rental value) of a legal interest in property. Valuers always prefer to use the comparative method for assessing market value or market rent, because it links directly to evidence of current market transactions. The other methods; profits, residual, contractor's and investment, are used when the comparative method cannot be used with full confidence because there is little, or no, evidence of comparable market transactions. When this is the case, the valuer needs to stand in the shoes of the most likely purchaser, or tenant, to simulate their thinking and the calculations they might carry out when assessing how much to pay for the property concerned.

The comparison method

The comparison method is used to value the main types of property, for example houses, shops, offices, and standard warehouses and factories. These are regularly sold or let in the market, giving plenty of evidence to support an assessment of the rental value or market value of similar properties.

The method can be broken down into three stages:

1. select market comparables
2. analyse comparable sales or rental data
3. adjust sale or rental prices for differences in location, condition, accommodation and market movements, then form an opinion of the market rent and/or market value of the subject property.

To use the method, valuers need to be fully aware of current economic conditions. Ideally, the market should be stable and there should be plenty of evidence of recent sales or lettings of similar legal interests, in similar sized properties, in the same condition and in the same area. Unfortunately, these ideal conditions rarely exist and the valuer needs to exercise professional judgment, based on experience of market sentiment, in order to adjust for differences between the comparable properties and the property being valued.

In selecting comparables, the aim is to find matching pairs. In other words, to find a property that is as similar as possible to the property to be valued and which has been sold or let on similar terms close to the date of the valuation. Normally, valuers prefer to rely on data they have collected from their own brokerage activities because they can obtain full information about the transaction. In most cases, due to the differing nature of each property, comparables will be less than perfect and adjustments will have to be made to bring the evidence into line with the subject property. These adjustments include adjustments for differences in the timing of transactions, location, condition and range of accommodation and for sale or letting conditions. Valuation by comparison is dealt with in more detail in Chapter 9.

The profits method

Where comparable rental transactions are not available, it may be necessary to adopt a method that recognises that property is normally capable of helping an enterprise earn a profit. This profit can be shared between remuneration for the business operator (or tenant) and rent for the property provider (or landlord). The rationale for the approach is grounded on the economic theory that the rental value of a property is a function of its earning capacity, productivity or profitability and rent represents a surplus. Economists call this idea the Ricardian theory of rent.

The profits method of valuation illustrates that the rental value of all land and property is derived from its earning capacity or productivity. The valuer is required to calculate rental value from the viewpoint of a potential occupier of the property. In contrast, the comparative method merely requires the valuer to study the outcome of negotiated transactions between several landlords and tenants. Nevertheless, it must be remembered that the tenants in these cases will probably have prepared a business plan which will include a calculation to check that the rent agreed can be paid out of the profits of the business.

The method is typically used to value business property which enjoys an element of *monopoly* and hence direct comparable evidence is not generally available. It is often used to value hotels, pubs, petrol filling stations, private nursing homes and leisure property, such as bingo halls, fitness centres and cinemas.

In outline, the profits method of valuation requires an estimate of the gross profits that can be earned by the business based on the audited accounts. All normal working expenses are then deducted, excluding any rent payments on the property. The resulting figure, known as the divisible balance, represents the amount available to be shared between the tenant, for remuneration for running the business, and the landlord, for rent for the property that accommodates the business.

If, however, there is some evidence of comparable transactions these may be used as a check on the figures produced by the profits method.

If the property is to be sold, the investment method will be used to convert the market rent found by the profits method into a market value (capital value).

The profits method is covered in Chapter 10.

The residual method

The residual method is used to value property with development potential, such as vacant land, or buildings that have potential to be refurbished or to have their current use changed to a more profitable one, for example where an old factory is to be converted into city-living apartments. Development potential exists when a property can be improved or developed so that its market value increases by more than the development costs, including an adequate profit for the developer.

Normally, developments tend to be unique, unless they are for standard forms of housing and industrial developments and, as a result, it is unlikely that there will be adequate comparable evidence. The only way to judge the market value of a property in its existing, undeveloped form is for the valuer to attempt to assess what the developer, who is likely to acquire the property, would be prepared to offer. The valuer needs to think like the developer and try to carry out the same calculations that the developer would make. The basic calculation is:

Land value = Gross development value
less
Costs of development, including developer's profit

A developer would use this equation to seek to identify the optimum form of development possible within the planning control system. This highest and best use development is the one that maximises the financial gap between the market value of the completed development, called the gross development value (often assessed using the comparison method and/or the investment method) and the total costs of development. The total costs of development need to include a developer's profit that is at least enough to make it worthwhile for the developer to invest time, enterprise and risk-taking skills to undertake the development. This financial gap is a surplus, or *residual* sum, available for the developer to spend on buying the property in its current undeveloped form, in competition with other possible developers. It indicates the possible market value of the property in its unimproved state, reflecting its development potential.

The weakness of the residual method is that the results depend on, and are very sensitive to, a wide range of input figures, which themselves are often difficult to assess with accuracy.

See Chapter 11 for more details of the residual method.

The contractor's method

The contractor's method, sometimes called the contractor's test, is a cost method of valuation. It is used to value property for its current use when the comparative, profits or investment methods cannot be used because there is no active market for the property being valued and there is no evidence of sales or letting transactions due to the specialised nature of the property.

Certain types of property are hardly ever bought or sold for their current use and, as a result, there is no comparative market evidence. They might be either specialist properties, for example industrial properties which are designed for particular manufacturing processes such as chemical works, research laboratories, airports or public sector buildings, such as schools, universities, public libraries, town halls, fire stations, hospitals and churches. These buildings are generally built by the organisation responsible for the provision of the service and there is no alternative organisation which requires the property.

When no market exists, *cost* may be a good indicator, or proxy, for value to the occupier or owner, including a potential owner. The method is based on the assumption that a potential buyer, in the transaction described in the Market Value definition, would not pay more to acquire the property than the cost of acquiring an equivalent new one. The economic concept of *opportunity cost* is of assistance in considering the logic of the method. Nevertheless, the approach is unreliable because market value is determined by the economic forces of supply and demand and not by the cost of production. The contractor's method is therefore sometimes referred to as *the method of last resort*, to be used only when other methods are inapplicable or impractical.

In summary, the technique involves assessing all the costs of providing a modern equivalent property, adjusted to reflect the age of the subject property.

Value of land and buildings = Value of the site
+
Adjusted replacement cost of the building

Chapter 12 considers the contractor's method in more detail.

The investment method

The investment method seeks to derive the market value of a freehold or leasehold interest in property from its potential to generate future income. Normally, all the main forms of property, such as offices, shops, warehouses, farms, houses and apartments, are capable of generating a rental income as they are regularly let to tenants who occupy as leaseholders. Landlords hold these properties as investments, in order to receive rents from the tenants. This gives them an investment return, which can be expressed as a rate of interest, on the capital sums of money they paid to purchase the legal interests in the properties.

Property has always offered investment opportunities for private individuals, builders, speculators, and property companies. As an investment, property is similar to many other types of investment where a capital sum is paid out in exchange for future benefits. In the case of property, the future benefits are income flows, in the form of future rent receipts, and future capital, in the form of lump sum receipts from the future sale of the property, which may or may not be greater than the capital sum paid out to acquire the property.

The method involves two main stages.

1. Analysis of comparable property sale transactions to establish the relationship between rental incomes and the capital prices recently paid by investors. The relationship may be expressed either as multiplier and/or as a yield. These figures are units of comparison.
2. Application of the results to the property to be valued to capitalise its rental income. This calculation converts the property's anticipated future rental income into its present capital value, indicating how much it can be expected that the property interest will sell for to an investor purchaser.

The patterns of anticipated future rental income and the level of risk associated with receiving them are very variable. As a result, property investment valuation methodology and the associated calculations range from simple multiplication sums to varied and extensive calculations.

Applications and different types of investment method valuation are considered in Chapters 13 to 17.

Summary of valuation methods

Determining which method to use and when can prove challenging to inexperienced valuers especially as, in many cases, alternative methods can be applied and in other cases a combination of methods might be used. This section summarises the application of methods and provides some examples where alternative methods and combined methods might be appropriate.

Different methods are applied to determine market rental and capital values.

Method	Capital Value	Rental Value
Comparison	yes	yes
Profits	no	yes
Residual	yes	no
Contractors	yes	no

The investment method allows rental values to be turned into capital values and vice versa.

The section concludes with an outline of the application of the five methods both individually and in combination.

- A vacant house for owner-occupation would in all circumstances be valued by the comparison method.
- A shop let on a lease would be valued using the investment method to find its capital value, although the comparison method would be used to find both the current rental value and the yield to be applied. In some circumstances, the rent might be found using the profits method.
- A site with planning permission for residential development might be valued by the residual method to find its capital value. However an alternative approach would be direct comparison, providing suitable comparable evidence is available. In ideal circumstances both methods would be used as a check. Direct comparison would be used to estimate the value of the completed development.
- A pub would be valued using the profits method to find the rental value and the investment method would be used to convert this rental value into a capital value.

- An owner-occupied factory or warehouse unit would be valued by direct comparison to find its rental value and then the investment method would be used to find the capital value.
- A nursing home would be valued by the profits method to find its rental value and the investment method to find the capital value. If sufficient evidence exists the comparison method can be applied using the price per bed or price per room as the unit of comparison.
- Land ripe for commercial redevelopment would be valued by the residual method although the gross development value will be found by a combination of the comparison method to find the rent and yield and the investment method to turn that rent into a capital value. The comparison method would be used, if possible, to check the figure found by the residual method.

Valuation layout

All valuations require some calculations, and the way these are laid out can be an important means of conveying information clearly and concisely, as well as minimising the risk of making a mistake. The following notes outline the conventional valuation layout. The examples of investment valuation calculations given in the later chapters of this book follow these conventions. The sample below is Example 14.9 from Chapter 14.

Valuation of freehold where the rent is subject to outgoings

Term			
Rent reserved IR terms		£4,800pa	
less freeholder's outgoings:			
External repairs say 10% market rent	£500pa		
Insurance say 2% market rent	£100pa		
Management say 10% rent reserved	£480pa		
		£1,080pa	
Rent paid on FRI terms		£3,720pa	
× YP 3 years at 9%		2.5313	£9,416
Reversion			
Market rent FRI terms		£5,000pa	
× YP in perp deferred 3 years at 9%		8.5798	£42,899
Market Value			£52,315
say			£52,000

> In my own firm, small though it was, I insisted on all valuations being made on coloured paper which was ruled with vertical lines on the right hand edge. This was intended to keep rental elements from being added into the column for capital sums — it also kept the date and the data well clear of one another when arriving at the answer.
>
> *Valuation: Principles into Practice*, Taylor V W, 5th ed, Rees W & Hayward R (Eds), Estates Gazette, 2000, p 758

The basic idea is that three columns on the right hand side of the page hold only numbers (very occasionally a fourth column is required). For handwritten valuations a narrow ruler, 3cm wide (cf 3.7cm for most rulers) is useful for drawing columns.

Words go in the text area to the left. Don't put words, lines to explanatory boxes or anything else (including numbers) into a column unless it is part of the calculation.

In simple investment method valuations the left hand side column is used for outgoings and deferring YPs, the middle column is used for rents and YPs, and the right hand side column is used for capital values.

It is a good idea to put 'pa', or whatever, after rents to avoid mistaking them for capital values. Rents for periods other than a year should be turned to annual amounts as soon as possible.

The assumption is that numbers in columns will be added. If subtracting write in the text area "less", if multiplying write 'x', and if dividing write 'divisor' (the latter after any words).

A single underline in a column means '='.

Each time a number is entered or a calculation is done move down a line, and consider whether the answer should move to the right.

Key issues are to keep rental and capital values apart, avoid losing some of the calculation and don't include something that should not be there.

Accuracy and rounding

When preparing a valuation it is normal to work to the nearest pound, disregarding pence.

Valuation tables usually give figures with four or more places of decimals. If the figures required are worked out on most pocket calculators the result will be displayed to eight or 10 significant figures, and considerably more significant figures can be produced if the calculation is done on a spreadsheet. While individual valuers will have different views on how many figures are shown and used, you can't go far wrong by using the whole number shown in valuation tables or worked out on a calculator or spreadsheet, but write down only four or five decimal places.

In *Abrahams* v *Ramji* (LON/00ANOCE/2007/0124) the Local Valuation Tribunal noted that the applicant's valuer had "rounded the YP to two decimal places and the PV to four decimal places whereas *Parry's Valuation and Investment Tables*, which are normally relied upon for enfranchisement valuations, provide four decimal places and seven decimal places respectively. This rounding will have had some effect on the valuation of the landlord's interest in the individual flats, and upon the total premium to be paid".

Because some readers will wish to work through the example valuations in this book we have normally done calculations using no more than four places of decimals and have rounded manually so that the calculations are internally consistent. However calculations in discounted cash flow tables have been performed using all the decimal places. In the few examples where we have worked on a different basis this is noted in the text.

The result of a calculation will rarely be a sensible looking round number. A valuation is an *estimate* and, for example, it would be silly to report that, "in our opinion, the value of the property is £3,500,654". It is therefore necessary to round the valuation figure to avoid the suggestion that the light bulbs have been included. A valuation should be rounded only once, at the end. This prevents cumulative rounding error. There is one exception to this rule: in asset valuations, using the Depreciated Replacement Cost (DRC) method, the values of the land and buildings should each be rounded before being added together, rather than rounding the final answer. (DRC is a cost method of valuation used when valuing specialised property assets, which are rarely if ever sold in the market, for financial reporting purposes such as company accounts). It is normal to round up or down by up to 1% to get to a sensible looking answer, unless the valuation is very low, in which case the valuation figure can be rounded to the nearest £50 or £250. A quick way of finding 1%, to decide on how much the answer should be rounded, is to cover up the last two digits. The word "say" is put in the text area to show that a valuation figure has been rounded.

9 The Comparison Method

"Although this sounds simple and straight forward, there may be many pitfalls for the unwary." *An Introduction to Property Valuation, Millington*, 5th ed, Estates Gazette, 2000.

This chapter focuses on the rationale, application and process of the comparison method. The remaining chapters of the book examine the other four methods in more detail. The profits, contractor's and residual methods are covered in Chapters 10 to 12, while the fifth method, the investment method, which is relatively complex, is explored in Chapters 13 to 18.

The comparison (or comparative) method of valuation is the most widely used of the traditional five methods, not only as a method in its own right but also as a component of each of the other methods. Millington's comment above also acts as a timely warning that the other noteworthy observation about this method is that in technical terms it is probably the easiest to grasp but, at the same time, arguably the most difficult to apply.

Procedure

If you wanted to buy a fridge or a pair of shoes, you would be faced with a potentially bewildering range of options. Although these are goods of very different types, as well as substantially different costs, the purchasing process undertaken for each would contain some common features.

Consider the purchase of a car. You might start by analysing your needs; the number of seats, fuel economy, body shape and so on. All of this has to be considered within the context of a more or less limited personal budget. This last constraint may reduce the choice somewhat but there are many different variables to be taken into account and many possible options. Final selection from this range of options is likely to be supported by some sort of analysis, which might include cost comparisons set against the perceived benefits of a particular make and model. This is a process of comparison and it is this process which underpins the comparison method of valuation.

The most common application of the comparison method of valuation is the residential market for vacant possession and the process of buying a house is, in essence, much the same as outlined above. Potential purchasers would look at the range of available options that meet both personal needs and pocket and undertake a process of comparison. One difference between cars and houses however is that houses tend to be more expensive than cars. The other essential difference though is that cars tend to be more homogenous than houses. In other words they are a more uniform product. Most cars have four wheels for example and, despite the range of types and models, they all perform a relatively simple basic function. In fact many will be almost identical.

The same cannot be said of a house. No two properties can be alike, if only because each one must occupy a different site. Even a pair of otherwise physically identical semi-detached houses in the same street will stand on different plots. So residential property is said to be heterogeneous. This means that the factors affecting value are many and varied and this in turn means that the process of comparing one house with another tends to require considerable knowledge and judgment. This process is illustrated in example 9.1.

Example 9.1

Number 31 Acacia Avenue, a three bedroom estate semi-detached house, is for sale. A prospective purchaser has researched the market and has found three comparable sales on the same street. These three properties have all sold in the last six months and the market has been relatively stable over this period. The house has space for a garage but no central heating.

The three comparable properties are more or less the same as the subject property but with minor variations. Comparable number one is 25 Acacia Avenue. This has been sold for £204,950. It has central heating but no garage. The second comparable is 27 Acacia Avenue which has a garage but no central heating. It was sold recently for £211,195. The final comparable is 29 Acacia Avenue. This property has a garage, central heating and a kitchen extension on the ground floor. Its sale price was £232,250.

Step 1: Summarise the available information

Property	CH	Garage	Other	Sale price
31 Acacia Avenue	×	×		Subject property
25 Acacia Avenue	✓	×		£204,950
27 Acacia Avenue	×	✓		£211,195
29 Acacia Avenue	✓	✓	Kitchen extension	£232,250

This is a very simple and straight forward summary. In more complex cases further columns can be added to reflect additional information which may have a bearing on analysis such as the date of sale for example. This type of table is sometimes referred to as a Schedule of Comparables, and can be a useful tool for summarising the available information about comparables and as the initial stage of analysis.

Step 2: Adjust each of the comparables in turn to reflect the physical differences between them and the subject property

No 25	Sale Price	£204,950	
Deduct central heating, say		£5,000	
			£199,950
No 27	Sale Price	£211,195	
Deduct garage, say		£10,000	£201,195
No 29	Sale Price	£232,250	
Deduct central heating, say	£5,000		
Garage, say	£10,000		
Extension	£15,000		
Total adjustment		£30,000	£202,250

Step 3: Make an assessment of the value of the subject property
At this stage the valuer would review all the evidence and use it to draw a final conclusion as to the value of the subject property. In this case the value would appear to be about £200,000.

A number of observations need to be made about this very simple example. First, it would be rare for three adjacent and similar properties to have been sold at the same time to provide such straight forward evidence of the value of the subject property. Second, property markets reflect large numbers of buyers and sellers many of whom will have imperfect information or may be under some sort of pressure to buy or sell. They do not always achieve the perfect bargain and so transaction prices do not necessarily fit into neat patterns. The evidence can often be very *lumpy*. This means that the valuer will need to look at all transactions with a critical eye and will often have to seek explanations for apparent anomalies in the market data. Sometimes evidence will have to be weighted or even dismissed altogether because not all evidence is of equal value. It should be noted as well that averaging the adjusted prices of the comparables is not considered to be good practice especially where there are large variations.

Adjustments for variations, like garage, central heating and extensions, can be made on a fairly objective basis. Although the valuer must not lose sight of the limitations of using costs as a measure of value, cost can be a reasonable starting point. So, if you bought a house without central heating, in most cases it would not be difficult to pay to have this installed and you could argue, therefore, that the installation cost would be a good measure. Even then it should be borne in mind that there might be some disruption involved in installation, but this should be set against the benefit of the opportunity to design a brand new heating system tailored to your own needs.

Other adjustments can be more subjective; open views for example or a larger than average garden. Such amenities may be more greatly prized by some purchasers, but even then it should be possible to make some sort of sensible judgment about how such differences would be viewed by potential buyers.

To summarise, the ideal conditions for providing good comparable evidence are properties which are physically similar and in the same locality where there have been a number of recent transactions within a relatively stable market. The following commentary summarises the main adjustments that may have to be made when considering comparable transactions.

Adjusting for time differences

The date of valuation and the date of the comparable sale must be noted. An assessment must be made of the market conditions at the time of the comparable property sale and at the date of valuation. Many valuers work with, act as, or are employed by, agents and are aware of the changes and trends in underlying market conditions. So, in a rising market comparables will be adjusted upwards, and in a falling market they will be adjusted downwards. In a stable market they will need no adjustment.

The experienced valuer should be able to make a simple percentage adjustment. Statistical records of price movements may help to support any adjustments made, although the valuer should bear in mind that national or regional statistics may not reflect what is happening in the local market. Due regard must be had to the definition of value and the required date of valuation. By definition, a market value assumes that a sale, following market exposure and negotiation, takes place on the date of valuation at the valuation figure.

Adjusting for differences in location

Unless the comparable is in the same location as the property to be valued, its usefulness will be questionable. The location must display the same characterises in terms of acceptability to prospective owners. In practice there can be very wide variations in value for similar properties in different locations.

Consideration of the factors that can affect value will indicate why this may be so. For example, proximity to airports, motorways, transport interchanges, places of entertainment, factories and petrol stations may have an adverse effect on residential values, but may have a beneficial effect on the value of commercial and industrial properties. Other factors include quality of access, level of security, available public services (such as broadband, gas and electricity), potential for flooding and level of soil contamination.

Adjusting for condition and accommodation

No two properties are ever identical, even if they match in floor area, as might be the case with apartments in a block of flats, they can differ in terms of distance to lift and stairs, outlook, and floor level. They may also differ in terms of the internal condition, quality of decoration, quality of fixtures and fittings, such as those in the kitchen and bathroom.

Adjustments may be necessary if there is a major difference between the condition of the property to be valued and that of the comparable property sold. Cost and value do not equate, so care must be taken to reflect the attitude that buyers in the market have towards good, average or poor condition. The difference in value is unlikely to equal the cost of putting right any lack of repair or structural defect. The value difference can be less than, or greater than, the cost of repair.

To allow for differences in size, it is necessary to use a unit of comparison. This could be expressed as a price per hectare or per square metre.

The more complex or unique the property, the more difficult it becomes to make adjustments for differences and the less likely it is for this method of valuation to be used. It could not, for example, be used to value specialist industrial properties.

Adjustment for sale conditions

When selecting comparables the valuer needs to ensure that the sales were truly market sales undertaken at arm's length. In practice, sales may occur because an owner is in a hurry to sell, possibly for financial reasons; and sales may be between members of the same family or between related company sales. All of these circumstances could mean that the sale price is probably not the market value.

Application

As Example 9.1 suggests, one of the main applications for the comparison method is the valuation of residential property for owner-occupation where the need for a reasonable number of transactions is normally relatively easy to satisfy. It used to be the case that for evidence of transactions valuers were reliant in the main on personal records and local knowledge. However these days there are a number of internet based sources providing free transaction data based on residential sale prices recorded by the Land Registry. These include "my property spy" (*http://www.mypropertyspy.co.uk*), "OurProperty" (*http://www.OurProperty.co.uk*) and "mouseprice" (*http://www.mouseprice.net*) and require a simple registration procedure to provide transaction and other information using a range of search criteria including post codes. The downsides to this type of data are that it tends to be out of date (it can take six months or more from the agreement of the purchase price for the sale to be completed, but it is only after completion that the transaction will appear in the public domain) and there is little detail about the property and the transaction.

This means that this type of valuation can appear to be rather backward looking and the comparison method is often criticised as being retrospective rather than forward looking. There is no guarantee that what has happened in the past is a guide to the future or even the present and uncritical application of the comparable method can lead to errors. One way to guard against this is to monitor actual transaction prices against asking prices and this can be a good guide to the current and future direction of the market. It has to be said that one of the skills required by the valuer is being alert to evidence of future market trends.

In addition to the valuation of owner-occupied residential property, the comparison method is widely used in the valuation of commercial and residential rents, agricultural land, and as a check on residual valuations of development land. In all these cases the procedure to be adopted can be reduced to the three basic steps outlined above.

1. Find recent comparable transactions.
2. Analyse those transactions to take account of variations
3. Weigh and adjust the evidence and draw conclusions.

It was suggested earlier that one of the challenges of the comparison method for the residential valuer is the lack of homogeneity of domestic property. Non-residential property does however tend to be less heterogeneous. In other words with non-domestic property, offices and factories for example, there are fewer variables. This means that analysis can be undertaken in terms of the unit price or by units of comparison. This is the subject of the next section.

Units of comparison

Size is one of the main reasons for variations in value between properties. The unit of comparison takes both the size and the value of a property into account.

Example 9.2

An office in a business park has a net internal floor area of 250m^2. It has just sold for £500,000. The analysis of this transaction would appear as follows:

Sale price divided by floor area = unit of comparison
or £500,000/250m^2 = £2,000/m^2

This data could then be used to value a similar business park office extending to 256 m^2:

256m^2 @ £2,000/m^2 = £512,000

Where size variations are very large, valuers need to take account of the tendency of the unit price to fall as the floor area increases. This phenomenon is called a quantum allowance (and is exactly the same as the general principle that if you buy more you expect a lower unit cost. Supermarkets, for example, will normally sell a dozen eggs at a lower price per egg than half a dozen eggs). There are exceptions to this general rule but the tendency is, in part, the result of the relatively low demand for large units. To take an extreme example; if you were trying to sell the Toyota motor manufacturing plant, located near Burnaston in Derbyshire, you would expect relatively few potential purchasers and the overall unit price is likely to be lower than the price applied to small starter units on modern industrial estates. A word of warning though, although the price per unit of comparison will tend to fall with scale, this is only a general rule and valuers should always be prepared to provide evidence to support making a quantum allowance in a particular valuation.

In the UK at least it would not be appropriate to use a unit price per area to value residential properties. However this is common practice in some countries, especially where the housing product is largely homogenous such as apartments in tower blocks.

Retail: Zoning

So, as a general rule residential property is valued by comparing overall sale prices and most non-domestic property is valued by reference to a unit of comparison. Retail property, however, represents a special case. Most shops are located in more or less busy locations with high pedestrian footfall. It is fairly logical, other things being equal, that the higher the pedestrian flow the higher the number of sales opportunities and the higher the value of the shop.

The special case of retail is illustrated by the following example. Consider the following three shops:

Figure 9.1 Three adjacent lock up shops

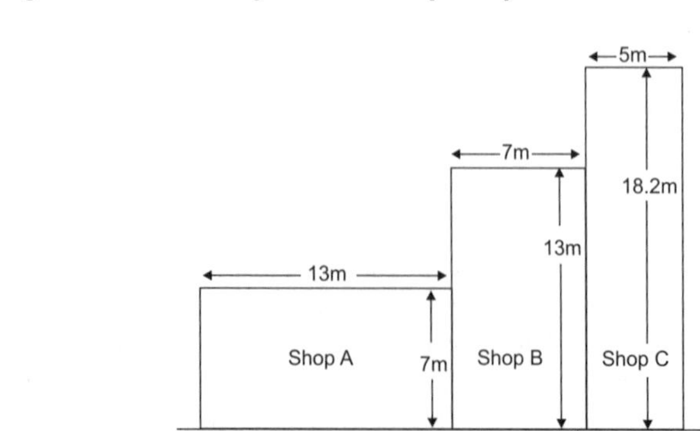

Shop A has a frontage of 13m and a depth of 7m. Shop B has a frontage of 7m and a depth of 13m. Shop C has a frontage of 5m and a depth of 18.2m. Shop C has just let at £75,000pa. What is the value of the other two shops?

The first step in the valuation is to analyse the available evidence (Shop C) and then apply the results of this analysis to value Shops A and B. The analysis, by a novice valuer, might be as follows:

Analysis Shop C

Frontage	Depth	Area
5m	× 18.2m	= 91m^2
£75,000pa	÷ 91m^2	= £824.17/m^{2}pa

Valuation of Shop A

Frontage	Depth	Area
13m	× 7m	= 91m^2
91m^2	× £824.17/m^{2}pa	= £75,000pa

Valuation of Shop B

Frontage	Depth	Area
7m	× 13m	= 91m^2
91m^2	× £824.17/m^{2}pa	= £75,000pa

This suggests that all three shops have the same value. However most people would readily regard Shop A as the most valuable owing to the longer frontage. In fact it is a widely accepted principle, when dealing with conventional shops in the High Street or in covered malls, that the greatest value is in the front portion of the shop. This is because the front of the shop is more accessible and the longer frontage allows more space for display and is more likely to attract customers.

This explains why variety shops, such as W H Smith, are organised in the way they are. Lines like newspapers and magazines, which are casual purchases and could be made from one of a number of outlets, are displayed at the front of the shop where they are easily accessible. The rear of the shop tends to be used for more specialised items such as books, CDs and videos. A shopper looking for the latest version of Grand Theft Auto is making a one-off purchase, and will be prepared to invest a little more time and effort than would be spent buying a newspaper.

The principle then is clear; the front of the shop is the most valuable, and this can be accounted for in practice by weighting the different parts of the shop in some way. We call this zoning. The shop is divided into a number of zones starting with the front zone or Zone

A. Zone depths do vary from place to place and may reflect historical oddities in town centre layouts. For example in older market towns, with narrow streets and long and narrow shop units, it might be appropriate to adopt a different pattern when compared with a modern shopping mall with more regularly shaped shops. However there is a tendency for Zone A depths in different locations to converge at something close to 6 m. (Valuers may encounter zone depths of 6.1m and although this seems odd at first, this is a conversion from the old imperial measure where zones were based on a depth of 20 ft). The next zone, Zone B, is usually measured as a further 6 m and, if necessary, a third zone will be introduced for any space to the rear of the shop. This is Zone C although it is sometimes referred to as the Remainder Zone for the obvious reason that it is what is left over.

Returning to our three High Street shops illustrated in Figure 9.1 we might apply these zoning principles and this is illustrated in Figures 9.2 to 9.4.

Figure 9.2 Shop A

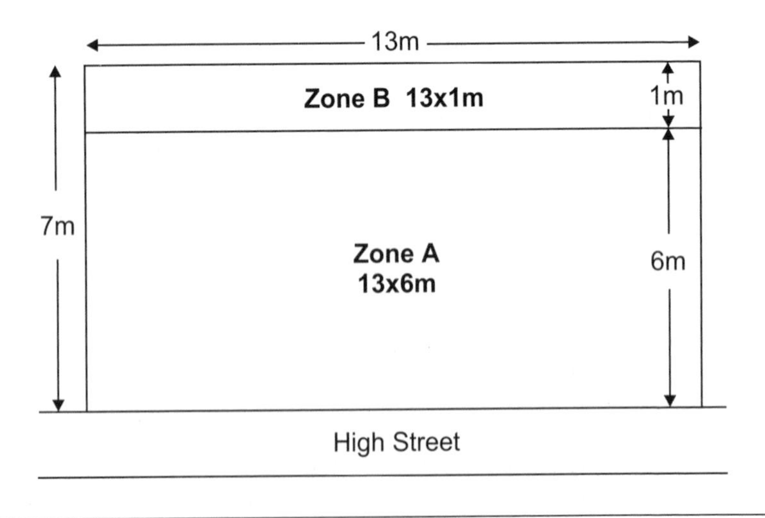

Figure 9.3 Shop B

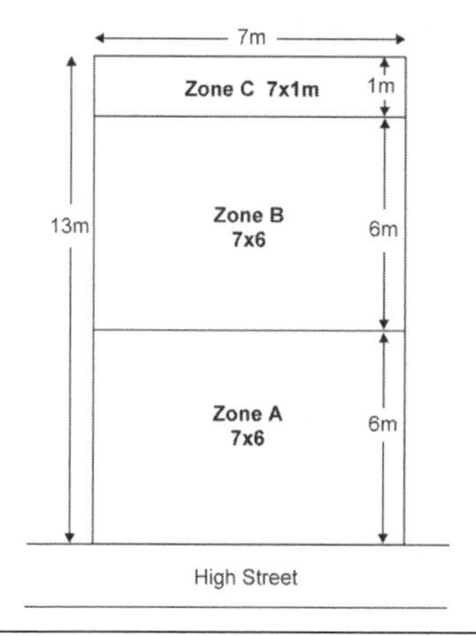

Figure 9.4 Shop C

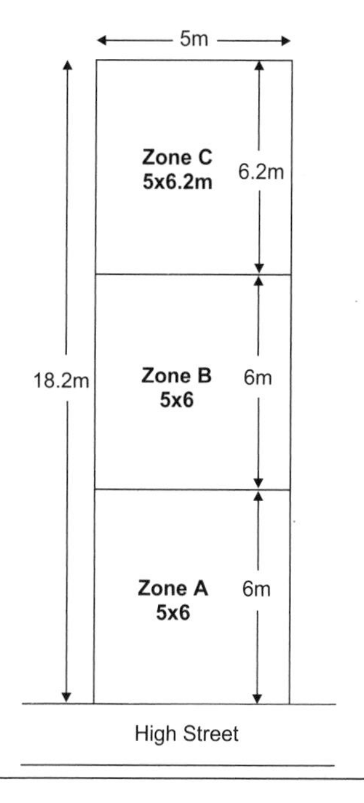

The analysis and valuation of these shops can now follow the zoning principle, whereby Zone B is weighted at half the value of Zone A and Zone C is valued at half the value of Zone B (ie 25% of Zone A). These adjustments may seem to be arbitrary but, as long as analysis and valuation follow the same principles (as you devalue so, shall you value), this should provide a sound result.

Analysis of Shop C

	Frontage		Depth		Weighting		Area
Zone A	5m	×	6m	×	100%	=	$30.00m^2$
Zone B	5m	×	6m	×	50%	=	$15.00m^2$
Zone C	5m	×	6.2m	×	25%	=	$7.75m^2$
Total Area in Terms of Zone A							$52.75m^2$

The resulting weighted floor area of $52.75m^2$ is usually referred to as the floor area in terms of zone A, often abbreviated to ITZA. To find the unit of comparison, that is the Zone A rent per m^2 pa of Shop C, we simply divide the rent by the weighted floor area:

£75,000pa ÷ $52.75m^2$ ITZA = £1,421.80/m^2pa ITZA

This price per unit area can then be applied in the valuation of shops A and B.

Valuation of Shop A

	Frontage		Depth		Weighting		Area
Zone A	13m	×	6m	×	100%	=	$78.00m^2$
Zone B	13m	×	1m	×	50%	=	$6.50m^2$
Total Area in Terms of Zone A							$84.50m^2$

Applying the value found by comparison the rental value is:

$84.50m^2$ ITZA × £1,421.80/m^2pa = £120,142pa, say £120,000pa

Valuation of Shop B

	Frontage	Depth	Weighting	Area
Zone A	7m	× 6m	× 100%	= 42.00m^2
Zone B	7m	× 6m	× 50%	= 21.00m^2
Zone C	7m	× 1m	× 25%	= 1.75m^2
Total Area in Terms of Zone A				64.75m^2

Applying the value found by comparison the rental value is:

64.75m^2 ITZA × £1,421.80/m^{2}pa = £92,062pa, say £92,000pa

Shop A turns out to be the most valuable, simply because more of the floor space is in Zone A. The relative values seem to be a fairer reflection of what a tenant would be prepared to pay than that produced by a straight un-weighted unit of comparison.

Shops come in all sorts of shapes and sizes including storage space, upper floors and basements. The zoning principle can be applied to all of these ancillary areas and will be based on local practice and the judgment of the valuer, always recognising that it is the overall value of the space that is being determined. So, basements and upper floors and storage areas may well be valued at as little as 10% or less of the Zone A price. This can be confusing at first but there are a few simple principles to be applied:

- sales space will have a higher value than non-sales areas such as storage
- for upper floors and basements used for sales, ease of access will increase value
- so, a first floor retail area with an escalator from the front of the shop will have a higher value than a first floor with access only by stairs at the back of the shop
- the value of ancillary areas are a small proportion of the value of the whole shop, and ancillary areas will have a very limited impact on the overall value.

The following table gives a guide to the percentages of Zone A which might be applied to other areas of a shop:

First floor sales/storage	15% to 20% depending on use and access
Second floor sales/storage	12.5% to 15%
Basement sales	20%
Basement storage	1% to 15%
Separate storage buildings	5% to 10%
Yards and outside storage	Nominal/spot figure

This should be seen as an indication of how valuers might approach the valuation of these ancillary areas but it would be wise to check local practice.

Another potential set of problems concerns shops that do not necessarily conform to our simple example. The following notes provide an outline of general approaches.

Return frontages

A shop with a return frontage is one with a street-corner position, providing a second area of window display and possibly a second doorway onto a side street. A return frontage may add value by improving the prominence of the unit to passing pedestrians. A simple solution is to reflect this enhancement by making an end allowance, adding a percentage at the end of the normal zoning calculation. End allowances are normally multiples of 2.5%, the size of the allowance depending on how much of an advantage the return frontage is. A return frontage onto a quiet side street will attract a small end allowance, while a return frontage onto a side street with a high retail rental value, and which has a significant foot fall, will be given a larger end allowance because tenants will be prepared to pay significantly more rent.

Disabilities

Some shops, particularly those in older and more traditional locations may exhibit a range of disabilities, such as changes in floor levels and obscured (or masked) areas not visible from the front of the shop. Every case needs to be treated on its merits but it may be that some adjustment to the normal zoning pattern, changing the depth of the zone to coincide with the change in floor level for example may well be appropriate. However in all such cases the valuer should consider the more robust solution of subtracting a percentage end allowance.

Small shops

Kiosks are very small shops often in busy locations, such as railway station forecourts. It is unlikely that the zoning approach will reflect the high pro rata value and the comparison approach may not be suitable at all. In such cases it might be appropriate to use a spot price per unit area or value the property by reference to its turnover or profits. This general method of valuation is explored further in Chapter 10.

Large shops

Large shops, such as department stores and supermarkets, will usually be valued on an overall price per m^2. Zoning is not normally appropriate because frontage and passing trade are not the most important determinants of value. This would also apply to retail warehouses.

Other units of comparison

The use of the comparable method is widespread across most classes of property. We have seen the unit of comparison for commercial properties is the rental per annum per m^2. The table below summarises a number of other possible applications.

Summary of units of comparison

Property type	Unit of comparison	Comments
Residential property for sale with vacant possession	Overall price — the valuer will need to distinguish between different types and sizes	Capital values
Commercial property for rent	Rent pa/m^2	Rental values
Retail units	Rent pa/m^2 in terms of Zone A	Rental values
Agricultural land	£/hectare or £/ha pa	Capital values or rental values
Development land	£/ha or £/m^2	Capital values. Used as a check on or alternative to the residual method
Hotels and care homes	£/bedroom	Used as a check on the profits method
Pubs	£/barrel	Used as a check on the profits method
Cinemas and theatres	£/seat	Used as a check on the profits method

Yield comparison

There is one final important application of the comparison method. We have seen how rental values of commercial, industrial and retail properties can be found by comparison but what of the market values, or capital values, of properties within these categories? Consider the following modern factory units on an industrial estate.

Example 9.3

Four Factory Units

Unit	*Rental value*	*Sale price*
Unit 1	£50,000pa	£525,000
Unit 2	£30,000pa	£295,000
Unit 3	£15,000pa	£165,000
Unit 4	£45,000pa	Unit to be valued

This approach requires an analysis of the relationship between the rental values and the sale price of each unit. Dividing the sale price by the rental value generates a multiplier. In essence this multiplier represents the number of years it will take for the rental income generated by the property to accumulate to equate to the sale price or, the number of years it will take (in effect) to purchase the property, and the multiplier is therefore called the years purchase.

Analysis

Unit	Rental value	Sale price	Sale price ÷ Rental value
Unit 1	£50,000pa	£525,000	10.50
Unit 2	£30,000pa	£295,000	9.83
Unit 3	£15,000pa	£165,000	11.00
Average			10.44

Valuation

The average multiplier is 10.44 times the rental value this could be used to value Unit 4. However, the averaging of the multiplier may distort the final answer and it is more appropriate for the valuer to use skill and judgment to decide on the multiplier. Unit 1 is perhaps the closest comparable in terms of size so that the multiplier adopted should be closer to 10.5:

Rental Value	£45,000pa
Multiplier	10.50
Capital Value	£472,500

As always it should be remembered that valuation is not an exact science and the valuer always needs to exercise a degree of judgment in assessing the relevance and quality of different comparables.

The application of the years purchase multiplier will be revisited in the context of investment method valuation (see Chapters 8 and 13).

We have seen then that the comparative method is a widely used method. It depends on evidence of recent transactions and the analysis of market data. Analysis (or devaluation) is the opposite of valuation but has to be done in the same terms: as the saying goes as you devalue, so shall you value. In all market sectors other than residential, units of comparison can be used to allow for variations in size. All other variables are allowed for subjectively and this means that the method requires skill and experience in its application.

10 The Profits Method

The profits method, which is sometimes referred to as the accounts method, the profits test or the receipts and expenditure method, is only used where it is not possible to value by comparison because reliable comparables are not available. This tends to be where some element of monopoly exists. The monopoly could be legal (for example licensing requirements for pubs, or the difficulty in getting planning permission for uses such as bingo halls) or factual, because no competing property is likely to be built, for example isolated trading locations like a book stall in a railway station or a garden centre in the countryside.

The profits method is used for, among other types of property, pubs, bars, restaurants, hotels, clubs, cinemas, theatres, bingo halls, petrol filling stations, some isolated retail properties, nursing and care homes, race courses, caravan sites and a wide range of leisure properties. This list is neither exhaustive nor fixed.

The argument behind the profits method is that the value of the land is related to its earning capacity. Because the tenant cannot build or lease alternative premises the landlord can demand a share of the profits as a rent payment.

There is a trend towards wider use of the profits method in assessing rents. Tenants see it as fair because they pay a share of what they make, while the landlord's income is secure because the rent is related to what the tenant can earn from the property. Both parties have an interest in the success of the business. Landlords will want to ensure that potential tenants have the skills necessary to run the business and pay the rent, and will usually insist that a tenant's application for a lease be supported by a well prepared business plan. A succession of revolving door tenants, whose stay in the premises is both short and unprofitable, will reduce the value of the property and damage the reputation of the landlord.

The profits method assumes that the hypothetical tenant trades solely from the property being valued. The profits method cannot be used if it is not possible to find the profit attributable to the individual property because the premises form an integral part of a wider business, for example a refinery operated by a major oil company (which would be valued using the contractor's basis — see Chapter 12).

The basic idea is shown in Figure 10.1 below.

Figure 10.1 The profits method

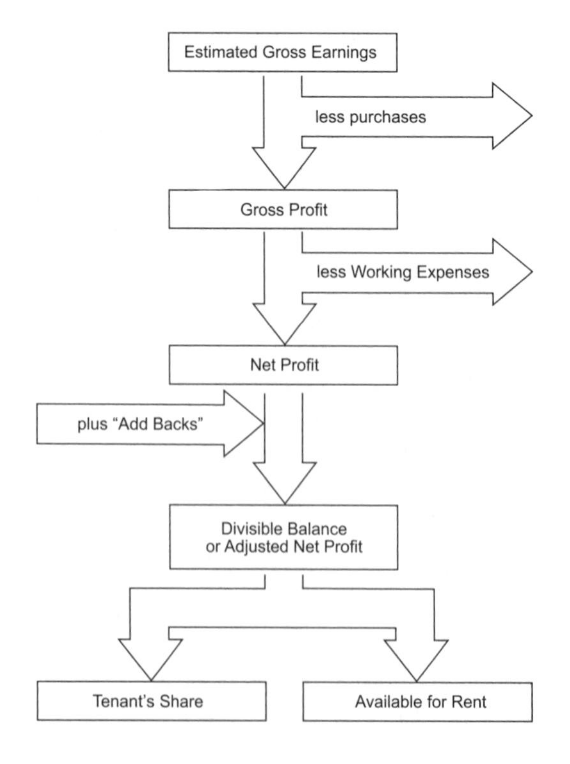

Ideally, the valuation is prepared using the audited accounts of the business trading from the premises. The source(s) of the information used should be stated in any valuation report.

It is important to consider a full year's figures because many businesses are seasonal and may, for example, have increased takings in the summer or in the run up to Christmas. If the accounts are for 14 months the valuer should take care to ensure that seasonal variations are allowed for and should not simply divide the figures by 14 and multiply by 12. The valuer should beware of accounts drawn up for periods other than a year. Six month accounts may indicate that the business is having financial problems and is being monitored closely by its bank.

If possible the valuer will look at three years' accounts. The valuer will normally work from the latest figures rather than taking an average, but will use the earlier accounts to get a view of how the business is trading and to even out fluctuations. Earlier accounts can also give an indication of trends. A year's accounts may be disregarded if they are affected by exceptional circumstances, such as a reduction in the number of tourists visiting the countryside due to an outbreak of foot and mouth disease.

If it is not possible to get the accounts the valuer's experience, judgment and skill come into play, and as a last resort estimated figures must be used. This would be the case, for example, where the valuer is required to value a new venture as part of the completed development value within a residual appraisal (see Chapter 11).

The following sections deal with each of the main elements of the profits method of valuation.

Estimated gross earnings

The estimated gross earnings are the takings or turnover of the business excluding Value Added Tax. The valuer has to find the fair maintainable level of trade (FMT) which would be achieved if the premises were occupied by a reasonably efficient operator, not necessarily the takings of the business operating from the premises.

The valuer should consider changes which will affect future takings. For example, the introduction of the smoking ban and the availability of cheap beer from supermarkets have significantly reduced the turnover of many pubs; a proposed road scheme would have a dramatic effect on the value of a petrol filling station if it limited access or meant that fewer cars passed the site; development in the area might increase the occupancy rate of a small hotel temporarily while builders are staying there, however the takings will be reduced when the development is finished and could fall further if a rival hotel was being built as part of the scheme.

The valuer should also consider if there is potential for profits to be increased if the tenant's fixtures, fittings and equipment were replaced, or if the business could be improved by better management or by moving into new lines, for example by a pub offering meals.

If accounts are available the valuer should look out for badly run businesses which will have correspondingly low profits, caused, for example, by restricted opening hours or by being run by an incompetent proprietor like Basil Fawlty. The figures will have to be adjusted so that the landlord does not subsidise a poor business operator.

The valuer should look out for fiddles. For pubs the lease will often be subject to a tie, requiring the licensee to sell only beer (and possibly wines, spirits and minerals) supplied by the landlord. It is not unknown for tenants to sneak in private supplies the sale of which will not appear in the accounts.

Some of the profit may be personal to the proprietor, for example a business run by a company or under a franchise which attracts extra customers with national advertising or simply by having a well known brand name, or a celebrity TV chef running a restaurant. This personal goodwill should be excluded from the calculation because it is associated with the proprietor of the business and would not be available to a hypothetical incoming tenant.

Purchases

'Purchases' is simply the cost of the goods bought in by the business for manufacturing and/or resale, again excluding VAT.

Where the premises are run as part of a group, bulk purchasing may reduce the cost of purchases, in these circumstances an allowance should be made to increase the cost of purchases to reflect any discounts obtained, and the valuer should check that the property being valued has not had an unfair share of the cost of purchases allocated to its accounts.

The *Estimated Gross Earnings* less the *Purchases* is the *Gross Profit*.

Working expenses

The working expenses include wages, services (electricity, gas, water, drainage, telephone and internet service provider), insurance, renewal of fixtures, fittings and equipment, advertising and business rates (unless valuing for rating) and any other costs to the tenant associated with running the business, but *not* rent. Rent is excluded because this is the figure that the calculation is trying to find, and rates are excluded in a rating valuation for the same reason.

Repairs should be included if they are the tenant's responsibility, but no allowance should be made for the cost of repairs for which the landlord is liable. A sinking fund may be included to replace some items as they wear out.

If the premises are run as part of a business with several branches it may be appropriate to make an allowance for head office expenses, such as marketing, training, accountancy and other management costs.

The *Gross Profit* less the *Working Expenses* is the *Net Profit*.

The net profit

The net profit is sometimes referred to as the EBITDA — the earnings before interest, tax, depreciation and amortisation.

It may be necessary to adjust the figure for net profit shown in the accounts to allow for:

a. the proprietor's salary, which may appear in the accounts as drawings or directors' remuneration. If the proprietor's wife is paid for nominal work to avoid tax this amount should be added back. If, however, the proprietor's wife works for the business but is not paid a proper wage a deduction should be made
b. interest payments, mortgage payments and any abnormal bank charges (the actual operator's financial position is irrelevant)
c. depreciation and
d. any private expenses put through the business.

These adjustments are called add backs.

The result is the *Divisible Balance* or *Adjusted Net Profit*, which is available to be split between the tenant and the landlord.

Tenant's share

The tenant's share represents the opportunity cost of the tenant's time and money invested in the business.

The tenant's share consists of:

a. interest on capital — a commercial rate of interest is taken on the value of plant and machinery, fixtures and fittings, equipment, vehicles, and working capital (stock, cash in hand and bank current account). For many businesses it is usual to estimate the working capital by assuming that the stock is three weeks purchases and allowing for cash in hand and in the bank at three weeks expenses. No allowance is made for the value of land and buildings and
b. tenant's remuneration a reasonable allowance for the work of the tenant (remember that the proprietor's salary/drawings were added back earlier to avoid double counting), including an amount to reflect risk and enterprise.

As a very rough guide the tenant's share is often about half the divisible balance, although if the market for the property is good the hypothetical tenant may be prepared to pay a higher proportion of the profit as rent, while if there is little demand the hypothetical tenant will pay less.

The tenant's share should also be considered as a monetary amount, to ensure that it is enough to provide a satisfactory livelihood for the proprietor and give a proper return for the tenant's efforts and for the risks being carried.

As an alternative to taking a proportion of the divisible balance the tenant's share can be estimated by taking a percentage of the gross earnings.

Amount available for rent

The balance after deducting the tenant's share is available to pay the landlord as rent. It may be necessary to adjust this figure to reflect any unusual terms in the lease, alternatively this can be allowed for by increasing the risk element in the tenant's share.

Example 10.1
Valuation of a small hotel

Estimated gross earnings (net of VAT)			
lettings		£90,000	
catering		£100,000	
bar takings		£80,000	
		£270,000	
less purchases (net of VAT)		£70,000	
Gross profit		£200,000	
less working expenses (net of VAT where appropriate):			
wages and national insurance	£54,000		
repair and renewal of fixtures, fittings and equipment	£2,600		
heating and lighting	£8,000		
rates	£20,000		
laundry and cleaning	£4,000		
water, licence and telephone	£2,750		
sundry expenses	£7,750		
		£99,100	
Net Profit (or divisible balance)			£100,900
less tenant's share:			
Interest on capital			
Fixtures, fittings and equipment		£36,000	
Cash (3 weeks expenses)		£4,563	
Stock (3 weeks purchases)		£4,038	
		£44,601	
@ say 6%		0.06	
		£2,676	
tenant's remuneration and risk, say		£50,000	
			£52,676
Rent			£48,224
Say			£48,000pa

The valuation should be checked by direct comparison if possible, either a price per unit area or some other unit of comparison, for example for cinemas and theatres £ per seat, for pubs £ per barrel, and for hotels and care homes £ per bed. It is important that the basis used for comparison is relevant, and that account is taken of location and trading circumstances.

The profits method finds a rental value. If a capital value is required the market rent found by the profits test can be capitalised using the investment method to calculate the capital value of the land and building (see Chapters 13 and 14). Alternatively, a multiplier can be applied to the net profit to produce the market value as a fully equipped operational entity having regard to trading potential, which includes the value of the business carried on in the premises in addition to the value of the land and buildings.

Many properties valued by the profits method require refitting or refurbishing on a regular basis. The valuer should reduce the capital value figure to reflect any significant costs likely to be incurred in the short term.

Issues with the profits method

1. There can be problems over the disclosure of the occupier's accounts. The lease may require the tenant to produce the necessary information, which will often be regarded as confidential (and therefore should not be quoted either in a report which may be made public or when providing details of comparables in any later valuation). The valuer must then make an assessment of how reliable the figures provided are. If there are doubts about the accuracy of the accounts the valuation report should recommend that the information be verified before the valuation is relied on or published, and that an opportunity be given to revise the report if there are any material discrepancies.

2. The valuer's judgment on the share of the profit that goes to the tenant has a critical effect on the figure for rent. In the example above if the hypothetical tenant was prepared to accept £45,000 for remuneration and risk the rent would go up by just over 10%.

   In *Garton* v *Hunter* (VO)(1969) 210 EG 769, 889 the Lands Tribunal commented "... The defect of the profits test is that the amount of the tenant's share, which is a key element in the calculation, is founded not on bedrock evidence, but only on the judgment of the valuer. Even so the margin for error is contained within acceptable limits ...".

3. While valuers cannot anticipate all future trading trends, they need to be aware of the history of the premises and general trends

in the business concerned, including changes in legislation which may affect trading potential.

4. The profits method is normally applied to properties which require specialist knowledge and skill. The valuer must be very familiar with the market for the type of property concerned, the likely purchasers and tenants, the anticipated levels of trade, current and likely future levels of competition, and any other risks involved.

5. The valuation report should include an assumption that all licences, consents and other approvals necessary for a business to continue to trade from the premises will be renewed.

Further reading

RICS Valuation Information Paper No 2, *The Capital and Rental Valuation of Restaurants, Bars, Public Houses and Nightclubs in England and Wales*, 2nd ed, 2006.

RICS Valuation Information Paper No 3, *The Capital and Rental Valuation of Petrol Filling Stations in England, Wales and Scotland*, 2003.

RICS Valuation Information Paper No 6, *The Capital and Rental Valuation of Hotels in the UK*, 2004.

RICS Valuation Information Paper No 11, *The Valuation and Appraisal of Private Care Home Properties in England, Wales and Scotland*, 2007.

11 The Residual Method

The residual method is used to assess the value of property which is suitable for development or re-development. The volume, type and density of development are likely to be unique to each case and so comparable evidence is often not available for such property. In these circumstances it is necessary to calculate the value of the property by using the figures that a hypothetical developer might use when deciding how much to spend acquiring it.

An undeveloped piece of land on which some development is anticipated would be a typical situation where the residual method would be used. Development includes the demolition of existing buildings and the construction of new buildings. It would also be used where there is an existing building which is to undergo some redevelopment, including major refurbishment.

A property will have development potential if its value for its current use is less than its value taking account of its potential to be developed. Potential for viable development, or re-development, will exist if a property's market value can be increased by more than the cost of development.

Two approaches: comparison and residual methods

There are two approaches to the valuation of property with development potential:

- using the comparison method, if there is evidence of sale prices of land for comparable development or
- modelling how a developer would assess the worth of the property, using the residual method.

Most developments are unique, so it is often not possible to use the comparison method. If however the valuer has comparable evidence of several recent transactions of similar property with the same development potential, it will be possible for the comparison approach to be used to arrive at a market valuation.

Comparable evidence may often exist for residential building land. For example, a building site for a house may sell for £100,000, a figure which reflects the development potential of the site which, with other similar evidence, provides a unit of comparison for similar single house plots. Larger residential land transactions, where the type of construction and density of development is similar, may be analysed on a price per hectare or price per plot basis, and this can be used as a unit of comparison to assess the value of similar development sites. This approach may also be adopted with industrial sites, where development is often of a standard form and transactions are relatively frequent.

However most property developments involve unique designs, density, uses and costs of development. No direct comparable evidence will exist and the residual method must be employed to simulate how hypothetical developers will react to the property.

The residual method

Where there is no adequate comparable evidence to indicate the market value of a development property, the valuer has to consider the worth of the development from the point of view of the most likely developer-purchaser of the property. The valuer aims to think in the same way that a developer would think. An experienced developer would assess the property market and decide on the optimum form of development. This will include thinking about what can be built, what the finished development will sell for, how much it will cost to build and how much profit is needed.

So, the valuer develops a financial model (or calculation) to predict market reaction to the exposure of the property on the market.

In its simplest form, the method can be expressed as a subtraction:

	Market value of completed development	£A
less	Development costs and developer's profit	£B
	Residual sum for purchase of property in existing form	£C

If the residual sum exceeds the value of the property in its current use, the proposed development will be viable. Provided the development is likely to secure planning consent and it produces the highest residual value, taking account of alternative forms of development, it will be the *highest* and *best use* of the property. This use represents the optimum, or most cost effective, development for which planning permission can be obtained and is likely to maximise the developer's profit. The residual sum on this basis will give a guide to the property's market value in its undeveloped form.

Although the residual method is a widely used method, it is acknowledged to be subject to substantial potential inaccuracy. It is important, therefore, that where the development is of a generally standard form, for example residential low rise houses or standard industrial units, comparable sale prices are used, where available, to check the valuation figure found by the residual method.

Example 11.1

A property located on a prosperous shopping street is currently used for storage purposes. For this (existing or current) use its market value is £1m. The freehold owner has recently obtained planning permission to convert the storage property into shops.

The following calculation shows the property's value, reflecting its development potential, from the viewpoint of a developer who is likely to purchase it:

	Value of completed development (ie shops)	£6m
less	Total development cost, including developer's profit	£2m
	Residual sum for purchase of property in existing form	£4m

The residual value, £4m, is available for the purchase of the property in its existing unimproved state. In the absence of other evidence, it is assumed that this surplus figure is the price that the property would fetch if exposed to the market.

So, £4m reflects its value in its highest and best use. This is the potential retail use which overrides the lower (existing or current) use value for storage of £1m.

In other situations, the residual calculation may show that development is not viable. For example, the residual calculation may produce a negative figure or a figure less than current use value, suggesting that development should not take place at the present time. It may be that at some time in the future the demand for property will produce an increase in the value of the completed development which in turn, provided costs do not increase as well, will generate a larger residual sum. This explains why some land and vacant buildings lie undeveloped for years. It should be noted that if a residual valuation produces a negative valuation figure this does not necessarily mean that the property is a liability with a negative value, merely that the proposed development scheme is not viable.

Example 11.2

This is a more detailed example: A freehold building, in a city centre shopping position, has planning permission for re-development behind its original frontage, which is listed as being of architectural interest. The development is designed to provide a large single retail unit of 400m^2 Net Internal Area (NIA) of zone A* equivalent floor space, together with 1,800m^2 NIA office floor space above the retail unit. From the date of acquisition, it is expected that building work will start after one year and the development will be completed, let and sold after three years. Net rents are expected to be £2,500pa per m^2 (Zone A) for the retail floor space and £300pa per m^2 for the office floor space. Building costs are expected to be £6,000,000, excluding fees. The valuer is instructed to advise a prospective developer on the price to offer for the property.

(*Zone A represents the high value, front of shop floor area, say 6m deep — see Chapter 9.)

Stage 1

Gross development value

(a)	Shop market rent (400m^2 × £2,500/m^{2}pa)	£1,000,000pa	
(b)	Office market rent (1800m^2 × £300/m^{2}pa)	£540,000pa	
	Total market rent	£1,540,000pa	
(c)	× YP in perp at 5%	20	
	Gross development value (GDV)		£30,800,000
(d)	less Disposal fees:		
	Agent's letting: say 10% of market rent	£154,000	
	Agent's investment sale: say 1% of GDV	£308,000	
	Lawyer's letting and sale: 0.5% of GDV	£154,000	
			£616,000
Net development value			£30,184,000

Stage 2
less Development costs

(e)	Building costs	£6,000,000	
(f)	Professional fees @ 15% of (e)	£900,000	
		£6,900,000	
(g)	Finance on building costs and fees @ say 12% = (e+f) × ([1.12]^1 – 1)	£828,000	
(h)	Marketing costs say	£75,000	
(i)	Contingencies @ 5% of (e+f+g+h)	£390,150	
(j)	Developer's profit @ 15% of GDV	£4,620,000	
	Total development costs		£12,813,150
	Gross residual		£17,370,850

Stage 3
Adjust for land acquisition costs and finance

(k)	Adjust for fees @ say 1.5% and Stamp Duty Land Tax @ 4%: total 5.5% = 1/(1+0.055) = × 0.9479		0.9479
			£16,465,829
	Adjust for finance on acquisition cost @ 12% over the 3 year development period: × PV in 3 years at 12% (see Chapter 13) = 1/(1.12)^3 = × 0.7118		0.7118
	Net residual land value		£11,720,377
	say		£11,720,000

Notes on Example 11.2

Stage 1 — Gross development value

(a) Current rental value figures should be used, rather than a speculative forecast of rental values at the time the building will be ready to let. The potential that rental value may be higher (or lower) in the future is reflected in the level of developer's profit required. The greater the potential for rental growth the lower the profit required. Selection of the appropriate Zone A net rental value has a big impact on the result but good comparable evidence for a non-standard retail unit may be limited, calling for skilled judgment based upon experience.

(b) Comparable evidence for offices is normally readily available.

(c) The all risks yield and YP are assumed to be those for which the building would sell if available for sale today (see Chapter 14, example 14.2). This avoids the problem of explicitly predicting the yield at the date of completion. The yield is selected by comparison with the analysed all risks yields of similar investment property. It should reflect the proposed property's investment qualities, the most important being its location and hence its rental growth potential. The greater the rental growth potential the higher the market value and the lower the all risks yield required by the market.

(d) Letting fees might be based on a scale of 10% of initial rent when one agent is employed or 15% when two agents are used. Letting fees are incurred at the end of the development period and so finance costs will not arise. Where the likely purchaser intends to sell the completed scheme, legal and surveyor's fees on the sale of 1.5% to 3% of gross development value should be deducted.

Stage 2 — Development costs

(e) Building cost is the payment to a builder for the actual construction. If appropriate, other costs may be included here, such as demolition of existing buildings, removing contamination, planning and building regulation fees, compensation to displaced tenants or to neighbours for right to light, the cost of any planning agreement with the local planning authority and any liability for Community Infrastructure Levy. Often there will not be sufficient time in preparing a residual valuation to seek detailed, expert advice on building cost. It is, therefore, common for the valuer to forecast construction cost on an appropriate cost per m^2 of gross internal floor area estimated from schematic drawings. There is no substitute, however, for direct experience obtained through analysis of contractors' tenders and final accounts. When in doubt, the valuer should refer to an appropriate expert (for example a quantity surveyor) and the source of the forecast should be made clear in the valuation report.

(f) Professional development fees can vary considerably depending on the nature of the work, the size of the scheme and the problems involved. Typically, they range from 5%, for simple residential estate development, to 15% for large complex multi-use development. The figure of 15%, in the example, is comprised of the following estimates: architect 7%, structural and other engineers 3%, project management 3% and quantity surveyor 2%.

(g) Finance charges on the total cost of development should be included in the budget. This is either the actual cost of borrowing or the opportunity cost of gaining a return on the capital from an alternative investment where the developer's own capital is used. The rate of interest is normally taken at 2% to 4% above London Inter-Bank Offer Rate (LIBOR), depending on the status and reputation of the developer. The figure of 12% is used here for illustrative purposes.

Normally, building contracts provide for the contractor to receive monthly payments based on the quantity surveyor's valuation of work carried out during the particular month. For the purposes of an initial appraisal of the development potential of a site, and in the absence of a quantity surveyor's cash flow forecast, it is common practice to assume that the monthly payments to the contractor will be the same throughout the building period. Since the developer will gradually build up borrowing to make payments as they become due, an estimate of finance costs (or interest paid) is often roughly calculated by assuming that the money is borrowed over half the building period. In this case, half of the two year building period is one year:

(£6,900,000 × Amount of £1 in 1 year @ 12%) – £6,900,000
= (£6,900,000 × 1.12^1) – £6,900,000
= £7,728,000 – £6,900,000
= £828,000

(See Chapter 13 for the formulae for the Amount of £1. In a case where the building period is three years, use the Amount of £1 in one and a half years, that is half of three years.)

(h) Marketing costs will cover advertising, letting and sale brochures and other promotional costs, incurred on an accumulating basis during the development period.

(i) A contingency allowance should normally be made for unforeseen increases in costs. The allowance may vary from say 2.5% for straightforward new construction to perhaps 10% for refurbishment schemes where unforeseen problems often arise. Alternatively, the valuer may reflect this aspect of the development process in an increased allowance for developer's profit (see j below).

(j) Developer's profit is an allowance for risk, enterprise, overheads and organisation. It is normally expressed as a percentage of the developer's capital expenditure. However, when preparing a residual land valuation, the full capital expenditure is not known, since the land cost (to be found) is part of the capital expenditure. It is necessary, therefore, to express developer's profit as a percentage of the gross development value (£30,800,000 in the example).

Although a 15% profit mark-up (£4,620,000 in the example) appears extremely generous to the developer, enterprise and risk must be rewarded. The allowance should be viewed both as a gross of tax profit and as one that must be kept at a minimum. This minimum figure should lead to a competitive bid for the site, while providing a reasonable cushion for the real possibility of rising costs. Inflating this figure will result in a lower bid than those offered by competitors but too low a profit margin would leave the developer exposed to risks such as falling values or increasing costs.

Stage 3 — Adjust for land acquisition costs and finance

(k) The gross residual figure includes the value of the land, the fees incurred acquiring the land, Stamp Duty Land Tax and the finance costs on land acquisition. To eliminate the finance costs the gross site value is multiplied by the present value of £1 at the finance rate of interest over the development period. The development period may be taken as the period between the date of purchasing the site and the date on which the completed property is let or sold. In the example, therefore, the development period of three years comprises one year, from the date of acquisition to the start of building, plus two years of building period. (If it is anticipated that letting and disposal of the completed development will be after the building period this additional time must be added to the development period). Acquisition costs have been taken as 5.5% (comprising surveyor's fees at say 1%, legal fees at say 0.5% and Stamp Duty Land Tax at 4%) of the net land cost.

Liability of the residual method to error

The residual method is subject to inaccuracy. One reason for this is that it is hard to provide accurate estimates of some of the costs of development. Building costs, for example, are set in a competitive market and contractors' bids typically exhibit a wide range. So, the valuer's first estimate of the value of the land may be substantially modified before a price is agreed. Acceptable profit levels for the developer are also subject to substantial variation. The valuer, therefore, must be aware that estimates of the various development costs may be different from the estimates used by competing buyers.

The difficulties of accurate selection of inputs to the residual calculation are compounded by the fact that the residual value is normally extremely sensitive to slight variations in the estimated inputs. The residual figure is particularly sensitive when it forms a small proportion of the value of the finished development, which is a result of subtracting total development costs that are close to the gross development value.

If we refer back to Example 11.1 the residual value is relatively insensitive to a change in the input figures. For example, if a revised calculation uses a 10% increase in the cost of development it will only produce a 5% decrease in the residual value.

	Calculation 1	Calculation 2	% change
Gross development value	£6.0m	£6.0m	0%
less Cost of development	£2.0m	£2.2m	+10%
Residue	£4.0m	£3.8m	–5%

If, however, costs of development had been close to the gross development value, as is normally the case, the residual value would have been very sensitive to change.

	Calculation 1	Calculation 2	% change
Gross development value	£6.0m	£6.0m	0%
less cost of development	£5.0m	£5.5m	+10%
Residue	£1.0m	£0.5m	–50%

Using a residual approach to find developer's profit

Once a development property has been purchased, or the purchase price is known before the site is acquired, the residual calculation can be used to assess how much the developer's profit will be, as a residue, on completion of the development.

In Example 11.2 (above), if the site was acquired for the price of £11,720,000, the profit should be £4,620,000 (see line j), as originally specified.

The following calculation is a re-arrangement of the residual valuation in Example 11.2 to find the developer's profit.

Gross development value			£30,800,000
less costs of letting and disposal			£616,000
Net development value			£30,184,000
less			
cost of site		£11,720,377	
Fees and stamp duty on site acquisition @ 5.5%		£644,621	
Total acquisition cost		£12,364,998	
Finance on acquisition costs — x Amount of £1 3 years 12%		1.4049	
		£17,371,586	
Building costs	£6,000,000		
Professional fees @ 15%	£900,000		
	£6,900,000		
Finance on building cost and fees For half building period × Amount of £1 in 1 year at 12%	1.12		
	£7,728,000		
Marketing costs say	£75,000		
	£7,803,000		
Contingencies @ 5%	£390,150		
		£8,193,150	

Total development costs	£25,564,736
Developer's profit	£4,619,264
say	£4,620,000

(Approximately 15% GDV)

Once the site is acquired the developer's profit is very sensitive to changes in the figures during the development period. For example, if during the development period market rents, and hence gross development value, increase by 5% compared to the original estimate, while building costs remain unchanged, the developer's profit will grow disproportionately from £4,620,000 to £6,130,000 (see below), an extra profit of £1,510,000, which is approximately a 33% increase!

Original gross development value — GDV	£30,800,000
add 5% increase in market value	£1,540,000
Increased gross development value	£32,340,000
less increased disposal fees	£646,800
Increased net development value	£31,693,200
less total development costs (as above)	£25,564,000
Developer's profit (revised)	£6,129,200
(approx. 19% of GDV)	
say	£6,130,000

If, however, the gross development value falls by 5%, because of falling market rents, the original developer's profit would fall from £4,620,000 to £3,100,000 (see below), that is a fall of £1,520,000 which is 33% fall in profit! This development profit sensitivity, to adverse changes in the constituent figures of the residual calculation, demonstrates the crisis situation faced by many developers when the 2007–08 property crash led to falls in market values of over 20% in one year and the prospect of further falls in 2008–09 and beyond.

Original gross development value — GDV	£30,800,000
less 5% decrease in market value	£1,540,000
Decreased gross development value	£29,260,000
less decreased disposal fees	£585,200
Decreased net development value	£28,674,800
less total development costs (as above)	£25,564,000
Development profit (revised)	£3,110,800
(approx. 11% of GDV)	
say	£3,100,000

During the development period, most figures in the residual calculation will prove to be different to those used in the original valuation made prior to buying the site. As shown above, the developer's profit is normally very sensitive to these changes, just as the original assessment of the site value was sensitive.

Spreadsheets allow monitoring of changes in the developer's profit caused by other changes in the input figures, for example changes in the all risks yield, changes in the interest rate charged on money borrowed, changes in the building costs and changes in the building or development period.

Further criticisms of the residual method

Although the comparison method depends on the skills and expertise of the valuer, it is directly related to market conditions and may well give a more accurate valuation result than the indirect residual method. Often, however, sufficient sales of comparable development properties are just not available and therefore the residual method must be used. Results produced are generally reasonably accurate when used in a genuine attempt to strike a bargain in a competitive open market. The method, however, is open to manipulation if it is used by a valuer seeking to establish what the market would pay in a hypothetical situation. For example, if a property was being compulsorily purchased the valuer acting for the person whose property is being acquired may tend to adopt an 'optimistic' view of the costs and benefits of the development, while the valuer for the acquiring authority might counter with a relatively 'pessimistic' view of the costs and benefits. The resulting views of market value could be widely different when they are formed outside the real market place, due to the sensitivity of the residual figure.

The following extracts from Lands Tribunal cases are indicative of the concerns that can arise.

> "... It is a feature of residual valuation that comparatively minor adjustments to the constituent figures can have a major effect on the result" ... and "... once valuers are let loose on residual valuations, however honest the valuers and however reasoned their arguments, they can prove almost anything". *First Garden City Ltd* v *Letchworth Garden City Corporation* (1966) 200 EG 163.

Nevertheless, in another case, where a residual calculation had been genuinely prepared as a basis for purchasing a property at an auction, the method was accepted as valid. The Tribunal commented that:

> "... it is a striking and unusual feature of a residual valuation that the validity of a site value arrived at by this method is dependent not so much on the accurate estimation of completed value and development costs, as on the achievement of a right balancing difference between the two. The achievement of this balance calls for delicate judgment, but in open market conditions the fact that the residual method is (on the evidence) the one commonly, or even usually, used for the valuation of development sites, shows that it is potentially a precision valuation instrument." *Clinker & Ash Ltd* v *Southern Gas Board* (1967) 203 EG 735.

Whenever possible, therefore, a valuation by direct comparison should be undertaken as an alternative to, or at least a check on, a residual valuation, albeit that differences in the nature of the sites (such as size, ground conditions, access, levels, drainage and development type) for which evidence is available may make the exercise difficult if not impossible.

Pessimistic/optimistic residual calculations

A discussed above, it is almost certain that several of the forecasted variable factors in the initial appraisal (for example rental income, yield, building cost, rate of interest on finance, building and letting periods) will differ from those that will actually occur during the development process.

It is argued, therefore, that present practice tends to depend too much upon the acceptance of well informed best estimates of the variable factors without an exploration or assessment of the range of possible and probable outcomes.

A useful, if basic, method of gauging the sensitivity of the residual surplus value to such outcomes is to prepare a pessimistic and an optimistic appraisal. This method of bracketing the residual value helps maintain an open minded view of the site's value, provides a warning against trusting the correctness of the result produced by the valuer's first estimate and indicates the level of risk involved in the scheme.

> "... Where uncertain market conditions or other variable factors could have a material impact on the valuation, it may be prudent to provide a

> sensitivity analysis to illustrate the effect that changes to these variables could have on the reported valuation. This will be particularly appropriate where a residual method has been used". *RICS Valuation Standards*, 6th ed, 2007, Guidance Note 5, Valuation Uncertainty, para 3.4.

Residual valuations can be re-expressed in the form of a discounted cash flow DCF (see Chapter 16). Costs are divided up (monthly, quarterly or yearly) and net cash flows calculated. Short term finance is allowed for in each period by the discounting process. This is very useful for complex phased schemes when parts are let or sold before the whole development is completed. Computer software is available to assist with the calculations and with sensitivity analysis of development proposals. As with any proprietary software it is vital that the valuer understands the calculation that the computer is performing, as there is a danger of rubbish in, rubbish out! Having said this, development appraisal software is a useful tool for experienced practitioners.

Spot values for hope value

There are cases where development potential is relatively uncertain.

Hope value exists when a property has a market value in excess of its existing use value, which reflects some prospect of a more valuable future use or development. It takes account of the uncertainty of the prospects, including the expected time needed for planning permission to be granted and for overcoming any other constraints to allow the more valuable use to be implemented.

The judgment of the Privy Council in the compulsory purchase case of *Mon Tresor & Mon Desert Ltd* v *Ministry of Housing and Lands* [2008] UKPC 31; [2008] 38 EG 140 stated that:

> "Where comparable sales are not available, resort may be had to the residual value method. This should be reserved for exceptional cases and will not be applied where the open market value is otherwise ascertainable by such assessments as a spot valuation. A spot valuation can take into account the existence and amount of hope value. Its assessment depends upon an amalgam of factors: the likelihood (ranging from complete certainty to a slight possibility) of the requisite planning permission being granted, the demand for the suggested development, the time that such development would take and the projected costs. The resulting figure represents the premium over existing use value that a

developer may be thought willing to pay in order to acquire the land in the hope of turning it to profitable account."

Further reading

RICS Valuation Information Paper Number 12, *Valuation of Development Land*, 2008.

12 Cost Based Valuation Methods

The contractor's method

Chapter 8 introduced methods of valuing properties which are usually sold or let on a regular basis in the property market. These are the *Comparative Method* and the *Investment Method*. It also explained that the Profits Method may be appropriate when property enjoys an element of monopoly and the *Residual Method* may be used when property has development potential. However, there are some types of property where these methods cannot be used reliably. The properties involved are usually non-profit making, publicly-owned buildings, or privately-owned specialised buildings which are rarely, if ever, sold or let in the market, because they are designed, used or located to meet the specific requirements of their owners.

Examples in the public sector include schools, universities, town halls, museums, police stations, fire stations, crematoria, swimming pools, sports stadia and churches; and in the private sector; oil refineries, chemical works, power stations, dock installations, breweries, glass works and cement works. The valuation of such properties may be required for a variety of purposes, such as business rates for local taxation, asset valuation for financial statements, for example company accounts, and compensation for the compulsory purchase of specialised property, for example churches and war memorials which need reinstatement following road widening. Here the valuation is hypothetical since the assessment of value for such purposes will not normally be put to the test of an actual letting or sale.

The rationale underpinning the contractor's method, sometimes called the contractor's test, is that, if the subject property was not available, the owner would have to acquire an alternative site and construct new buildings to continue to provide the service in the public sector, or to keep the business operating in the private sector. This cost, therefore, represents the maximum amount the user would be prepared to pay for the provision of the facility.

A cost approach to valuation has a theoretical weakness. It is founded on the fallacy that cost is equal to value. The number of bankrupt builders and property developers at times of economic recession illustrates the fact that land cost plus building cost does not necessarily equate to value, because value is determined by the interaction of the forces of supply and demand, not by the cost of production.

When a market does not exist, however, these forces cannot operate and cost may be a rough indicator of value to the occupier, including a potential occupier. This then is one circumstance where cost can be used as a proxy or substitute for value.

Generally, the order of preference when adopting a method for assessing value is:

1. the comparison method
2. the profits, investment or residual methods (as appropriate)
3. a cost based approach.

So, although the cost approach is the method of last resort, it is still frequently employed and certainly is to be regarded as an established method for certain types of property, despite its theoretical faults.

Different names are given to different cost assessments. The Contractor's Method, is associated with valuation for business rates and with the assessment of compensation for compulsory purchase on an equivalent reinstatement basis under Rule 5, section 5, Land Compensation Act 1961. It forms the basis of the Depreciated Replacement Cost Method (DRC) which is used for asset valuations for financial statements such as company accounts. The detail of these valuations is beyond the scope of this book, and readers interested in pursuing the subject further are referred to the further reading list at the end of this chapter.

The five stages to a contractor's method valuation are set out below.

Stage 1
Find the estimated replacement cost of a modern equivalent building, including incidental costs, fees and finance. It is assumed that the site does not have any features that would increase the construction cost. Architectural excess in the subject property will not be replicated in the modern equivalent building unless a building with a high specification and distinctive design would be required. Thus a Victorian town hall might be replaced with a modern office building, while a national parliament would require a grand design.

Stage 2
Adjust the replacement cost to reflect the disadvantages of the actual building, including physical, functional and economic obsolescence, poor design, planning and layout.

The depreciation factor might be calculated by dividing the future economic life of the building by its total life expectancy:

$$\text{Depreciation factor} = \frac{\text{Estimated future life of the building}}{\text{Total life of the building (from new)}}$$

Alternatively the valuer's skill and judgment may come into play to decide on the appropriate adjustment.

The result is the adjusted replacement cost or depreciated replacement cost of the building.

Stage 3
Add the capital value of the site, or an equivalent site, found by comparison. The result is the capital value of the property.

Stage 4
If a rental value is required for rating, the capital value is decapitalised by multiplying by an appropriate rate of interest cost to arrive at a pseudo-market rental value. Rating legislation determines the statutory decapitalisation rate. There are usually two fixed rates for a given rating list, for example 5% for general purposes and 3.3% for special circumstances where there is a moral obligation to provide the service such as health and education.

Stage 5
Stand back and look. At this stage the previous four stages are reviewed to see whether they produce a final figure that the hypothetical tenant

or the owner would pay. Where the adjustments cannot be accommodated in the previous stages they may be made in the form of an end allowance. This allowance might reflect the hypothetical tenant's ability to pay or other higgling of the market not capable of accurate reflection elsewhere in the calculation, such as problems with access to the site. The final figure is then rounded.

Example 12.1

You are instructed to value a local authority school of 1,306m², approximately 70 years old with a life expectancy of 20 years. The site area is 1.5 ha, 0.4 ha of the site is developed, the remainder is playing fields.

The estimated construction cost of a new building, including building cost, site works, professional fees and finance is £1,335/m². Development land in the neighbourhood sells for £2 million per ha, while land suitable for playing fields is worth say £20,000 per ha.

Valuation of School

Estimated replacement cost		
1,306m² @ £1,335/m²		£1,743,510
allow for obsolescence and depreciation		
@ say × 20/90		0.2222
Adjusted replacement cost		£387,408
add capital value of site:		
developed area 0.4ha @ £2 million/ha	£800,000	
playing fields 1.1ha @ £20,000/ha	£22,000	
		£822,000
Capital value of school		£1,209,408
say		£1,200,000
If hypothetical rental value was required the		
capital value can be decapitalised @ say 3.3%		0.033
Rateable value		£39,600pa

The contractor's method is not suitable if a valuation is required for secured lending purposes, such as a mortgage, because specialised properties do not normally have a market value, except perhaps for redevelopment purposes.

It is important to distinguish between the replacement cost of buildings when valued using the contractor's method and their reinstatement cost for insurance purposes.

Building reinstatement cost basis for insurance purposes

Assessment of building reinstatement cost for insurance purposes is another example of an assessment normally based on cost rather than market value. A reinstatement cost assessment is very different from a contractor's method valuation.

Building insurance is designed to provide the owner or occupier with protection against the effects of physical damage to the fabric of buildings from various risks or perils, defined by the insurance policy. These might be limited to fire, lightning and explosion but the scope of cover normally available extends to include damage from such events as riot, flood, burst pipes and subsidence.

In addition, it is normal to take out insurance to cover items such as consequential loss, for example loss of rent or profit, and third party liability, for example damages suffered by others due to failure of the fabric or negligence of the owner or occupier's staff.

The RICS *Guide To Carrying Out Reinstatement Cost Assessment*, 1999 provides useful advice.

Example 12.2

A commercial building reinstatement cost assessment:

Client: XYZ Ltd — Assessment Date: 15/09/08

Property: Albion Works, Sidney Street, S1 1WB

		Rate £ per m^2 x	Area m^2	£
1	*Buildings*:			
	Office	1,000 x	1,600	1,600,000
	Workshop	800 x	1,000	800,000
	Warehouse	400 x	10,000	4,000,000
	Disused boiler room	600 x	100	60,000
	Car park	60 x	3,000	180,000
2	*Add Special factors*:			
	Sprinkler system			10,000
	Marble finish to front elevation of office			20,000
3	*Deduct for Special factors*:			
	Boiler room roof — removed to limit business rate tax liability			4,000

4	*Site factors*:	
	add for difficult working conditions	
	— sloping site @ 3%	199,980
	add for neighbouring housing requiring	
	clean/quiet working @ 3%	199,980
5	*Location factors*:	
	add for restricted site access @ 2%	133,320
6	*Listing/conservation factor*: none	0
7	*Regional factor*: none	0
	Sub total	**7,199,280**
8	*Other allowances*:	
(a)	Asbestos/deleterious material clearance — cost	50,000
(b)	Demolition, temporary works, site clearance @ 5%	360,364
(c)	Allowance for Landfill Tax	5,000
	Sub total	*7,614,644*
9(a)	*Professional fees*:	
	Surveyors, architect, engineers,	
	project management etc @ 15%	1,143,397
(b)	Local authority fees:	
	Planning @ 0.5%	38,113
	Building Regulation @ 0.5%	38,113
Total		**8,834,267**
But say		**8,835,000**
Rebuild period = 2 years		

Cost information base used: BCIS

The assessment is made on the basis of total loss, requiring the entire building to be demolished and rebuilt. The calculation is to replicate the buildings as they exist, not to replace them with modern structures (as with the depreciated replacement cost and contractor's methods). The site cost is not included because the land will still exist if the buildings are destroyed. The effects of inflation and building cost increases are ignored but inflation is provided for within the insurance policy. Evidence of building costs can be obtained from various sources, such as BCIS, the RICS' Building Cost Information Service, which provides cost information to the construction industry and others who need comprehensive, accurate and independent data.

Most claims involve partial damage to a building rather than total loss and, therefore, they are normally settled on a partial reinstatement cost basis, requiring building work in the existing style and materials. So, except where there are no constraints over the nature of rebuilding, the cost of a modern substitute building will not provide the current basis for the assessment. This contrasts with the approach adopted in the contractor's basis described above.

Exceptionally, an insurance valuation may be on a market value basis. This may be appropriate where there is no intention of reinstating anything beyond comparatively minor damage or where equivalent property is readily available in the market at a figure well below rebuilding costs. If this is done it must be with the knowledge and consent of the client and insurance company.

It may be that this type of work is outside the scope of the valuation surveyor. In the case of a new building the quantity surveyor for the development is likely to be the most competent to assess the insurance figure. With older buildings both building surveyors and quantity surveyors are likely to be well equipped to undertake the assessment. For straight forward residential or simple commercial or industrial properties a valuer is likely to be able to make an assessment based on guidance provided by BCIS.

Example 12.3

The following example is a simple residential property insurance assessment, based on the house rebuilding cost calculator, provided by the Association of British Insurers (ABI) and the Building Cost Information Service (BCIS) at *http://abi.bcis.co.uk*:

BCIS House Rebuilding Costs

Base date:	November 2008
External floor area:	145m^2
Type of property:	Detached house
Age of property:	1920–1945
Regional Group:	South West
Rebuilding cost from ABI table:	£1,202 per m^2
Rate multiplied by area:	£174,290
Adjust for inflation from January 2008 to November 2008:	£9,309
Approximate rebuilding cost of house:	£184,000
Addition for garage:	£15,000
Addition for other items:	£2,000
Total:	£201,000

This on-line calculator gives a general indication of rebuilding costs for many common house properties within the UK, but it is not appropriate for all houses and the rebuilding cost of even similar houses can vary depending on individual circumstances. Subscription to the full BCIS service allows a more detailed, property specific, assessment.

Further reading

RICS Guide to Carrying Out Reinstatement Cost Assessments, RICS Books, 1999.

RICS Valuation Information Paper No 10, *The DRC Method of Valuation for Financial Reporting*, RICS Books, 2007.

Valuation and Sale of Residential Property, Mackmin D, 3rd ed, EG Books, 2007, particularly Chapter 15 on insurance valuation.

Valuation: Principles into Practice, Hayward R (ed) EG Books, 2008, particularly Chapter 9 study 2 on equivalent reinstatement, Chapter 12 studies 17, 18 and 19 on the contractor's test for rating and Chapter 16 study 3 for depreciated replacement cost for asset valuations.

13 The Time Value of Money and Valuation Tables

Before exploring the application of the investment method of valuation in the remaining chapters of this book it is necessary to understand the different ways of holding investment assets and the relationships between them. In essence all investment assets will be held in one or more of the following forms: *Income, Capital Now* or *Future Capital*.

Investment mathematics is about converting from one form of holding an asset to another. So, an investor might hold capital now and wish to convert this into an income. One way of doing this might be to buy a house and then let it to a tenant. Investment mathematics can be used to undertake such a conversion.

If there are three different types of financial asset, it follows that there are six ways of converting from one asset to another and there are six separate formulae governing these conversions.

All conversions are based on the fundamental principle that money effectively reduces in value over time. It is important not to confuse this idea with the depreciatory effects of inflation for, although this is connected, the time value of money is a much more complex concept that is largely a function of human behaviour.

All the formulae are based on the rules governing compound interest and all are concerned either with the process of compounding or its opposite, discounting. The compound interest formula is known as the Amount of £1. Its opposite, the discounting formula, is known as the Present Value of £1. Between them these two formulae govern the conversion between capital now and future capital.

The other four formulae are all concerned with the relationships between capital and income. The Years Purchase (or Present Value of £1 pa) converts an income flow into capital now. Its opposite, The Annuity £1 will purchase (or Annual Equivalent of £1) converts a sum of capital now into an income.

The final pair of formulae convert income to, and from, future capital. The Amount of £1 pa converts an income flow into a future capital sum. Its opposite, The Annual Sinking Fund, converts a future capital sum into an income flow.

Mastery of the six formulae is an essential prerequisite to a sound understanding of the investment method of valuation. The formulae will work with any currency, not just pounds sterling.

The time value of money

Generally speaking, most people prefer to have money now rather than wait for it in the future. If your employer suggested that you should be paid annually in arrears rather than normal monthly payments you would be very reluctant to accept such a deal. There are a number of reasons for this. First, you need the money now to pay your rent or mortgage, to buy food and pay for fuel and so on. These are basic needs which cannot be deferred. Second, you might be concerned about the eroding effects of inflation — by the time you receive the money its purchasing power may be significantly lowered. Third, you could put the money in a deposit account as it is received and earn interest on it. Finally, you might not trust your employer. You have done your year's work but can you be sure that the employer will pay up?

So, this is about choice, it is also about risk and reward. If you are to be persuaded to wait for your money, you will only do this if you can be adequately rewarded for the additional risk, the loss of choice and the loss of interest. For example, you might find it easier to accept delayed payment if there is some additional benefit, such as an agreement to pay you an additional 50% on top of your normal salary.

From this analysis two basic principles emerge: people prefer not to wait for money but might be persuaded to do this if sufficient reward is made available. So, for a sum of money in the future to be equivalent to a sum of money now, the future sum must be larger. The difference will be a function of the length of time in the future and the amount of risk (real and perceived) attached to the delay. The risk is reflected in the rate of interest.

Compound interest

One of the simplest forms of investment would be a savings account in a bank or building society. The saver deposits an amount of money, leaves it in a savings account for a period of time and would expect to withdraw the original sum plus interest at some later date. Such an investor is taking a sum of *capital now* and converting it to *future capital*.

So, £1,000 invested in a building society at 5% pa compound interest would accumulate to £1,157 in three years time, as shown below:

	Capital at start of period	Interest	Capital at end of period
Year 1	£1,000	£50	£1,050
Year 2	£1,050	£52.50	£1,102.50
Year 3	£1,102.50	£55.125	£1,157.625

Interest accumulates both on the initial capital sum invested and on the interest paid each year. This payment of interest on interest is known as compounding.

The calculation is fine but is rather tedious, the more so for longer periods of time. However, it can easily be reduced to a formula:

$$£1{,}000 \times (1 + 0.05) \times (1 + 0.05) \times (1 + 0.05) = £1{,}157.625$$

(Multiplying by 1.05 adds 5% interest each year. Notice that the rate of interest, 5%, is shown as a decimal, 5/100 = 0.05).

Which in turn can be reduced to:

$$£1{,}000 \times (1 + 0.05)^3$$

From this we can derive a general formula for calculating the amount that a lump sum will accumulate to if invested at compound interest:

$$P \times (1+i)^n$$

Where:

P = is the capital invested or principal sum;
i = the rate of interest expressed as a decimal; and
n = is the number of years.

In practice, the principal sum is assumed to be £1 so that the formula can be written simply as $(1+i)^n$. This is used in most valuation texts although occasionally different annotations will be used in different disciplines.

This means that £1 invested now will grow to £1.157625 in three years if interest is payable at 5% pa and £1,000 will grow to £1,157.625, that is 1,000 times as much.

The Amount of £1

The compound interest formula is referred to as *the Amount of £1*, usually shortened to A, which is defined as the amount to which £1 will accumulate at a given rate of compound interest (i) over a given number of years (n):

$$A = (1 + i)^n$$

The Amount of £1 allows us to calculate the equivalence between *capital now* and *future capital*.

In Chapter 6 it was noted that between 1950 and 2005 the price of an average house went up from £1,940 to £190,760. The average annual increase in value can be found using the Amount of £1 formula:

Average house price in 1950 × Amount of £1 for 55 years = Average house price in 2005

£1,940 × A for 55 years = £190,760
A for 55 years = £190,760/£1,940 = 98.329897
$(1 + i)^{55} = 98.329897$
$1 + i = \sqrt[55]{98.329897}$
$1 + i = 1.087$
$i = 0.087$, and the percentage rate of increase is 8.7% pa.

Discounting

This is the opposite of compounding and it is used to convert *future capital* sums into their equivalent capital value today (*capital now*). Imagine you want to set aside a sum of money that can be used to replace your car in three years time. You estimate that the cost in three years time will be £10,000. How much would you need to invest in your savings account now, assuming interest is paid 5% pa?

If discounting is the opposite of compounding, it seems reasonable to expect that the formula should be the opposite as well. In maths, the opposite of a formula is its inverse, and the inverse of a mathematical expression is found by taking the reciprocal, that is dividing into 1.

So, if compounding is $(1+i)^n$ and discounting is the opposite of compounding, then discounting = $1/(1+i)^n$.

The Present Value of £1

The discounting formula is known as *The Present Value of £1*, usually shortened to PV, which is defined as the single amount needed to be invested to grow to £1 in a given number of years (n) assuming a given rate of interest (i).

$$PV = \frac{1}{(1 + i)^n}$$

So, to save £10,000 in three years at 5% pa we can substitute the values for i and n in the above equation:

$$1/(1 + 0.05)^3 = 0.8638$$

In other words, every 86 pence invested now will grow to £1 in three years at 5% pa compound interest. Or, to put in another way, every sum of £1 in three years is worth 86 pence now.

To grow to £10,000, we would need to invest £10,000 × 0.8638 = £8,638.

The next pair of formulae deal with the relationship between future capital and income.

Amount of £1 pa

A formula containing the words "per annum" is an indication that income is part of the conversion relationship. *The Amount of £1 pa* is the amount to which a series of investments of £1, made at the end of each year, will grow at a given rate of interest (i) for a given number of years (n). This formula is similar to the Amount of £1 formula, but in the case of The Amount of £1 pa the formula is concerned with converting a series of *income* flows into *future* capital. It is known as APA or Apa for short and the formula is as follows:

$$APA = \frac{A - 1}{i}$$

which can be expanded to $((1+i)^n - 1)/i$

A typical example of the use of the APA formula would be where it is necessary to calculate how much will accumulate if a series of equal regular savings are made in an account with a financial institution which guarantees a fixed rate of interest. For example, an investor will receive a guaranteed 8% pa provided they are able to save £10,000 each year over the next four years. What will the investment be worth at the end of the four year period?

Amount of £1 pa = $(A - 1)/i$
and $A = (1 + i)^n$
Substituting the i and n values: $A = (1 + 0.08)^4$
So, $A = 1.36049$
Substituting the value of A back into the APA formula:
$APA = (A - 1)/i$
$APA = (1.36049 - 1)/0.08$
$APA = 4.5061$
$£10,000 \times 4.5061 = £45,061$.
So, the value of the investment will grow to £45,061.

Annual Sinking Fund

The second of this pair of formulae is the Annual Sinking Fund, abbreviated to ASF or SF. This enables a *future capital* sum to be converted to an *income* flow. ASF is the annual investment that needs to be made at the end of each year to grow to produce a future capital sum of £1, at a given rate of interest (i) in a given period of years (n). This is sometimes used to estimate the annual amount that needs to be set aside, out of income, to meet some future capital expenditure. Consider the following example:

You would like to know how much you would need to set aside each year to accumulate to £10,000 for a replacement boiler given that you can invest these annual sums over five years at a rate of interest of 3% pa.

If ASF is the opposite of the *Amount of £1 pa* then the ASF formula is the inverse of the APA formula:

$$ASF = \frac{i}{A - 1}$$

and this can be expanded to $ASF = i/((1 + i)^n - 1)$.
Substituting i and n values the formula becomes:

$$ASF = 0.03/((1 + 0.03)^5 - 1)$$
$$ASF = 0.1884$$

Capital sum to be replaced	£10,000
× ASF to replace £1 in 5 years @ 3%	0.1884
Amount required for sinking fund	£1,884pa

Sinking funds are common in the property context and often form part of service charges attached to leases to spread the cost of anticipated future capital expenditure on major repairs and replacements over a number of years.

This leaves the final relationship and the last pair of formulae. *The Annuity £1 will purchase* converts *capital now* to *income* (we will see that this formula, among other things, can be used to calculate mortgage repayments), and The Years Purchase, also known as *The Present Value of £1 pa*, performs the important function in property investment valuation of converting *income* into its *capital now* equivalent.

Years Purchase (The Present Value of £1 pa)

This formulae is concerned with the relationship between a property's rental income and its capital value. *The Years Purchase* is the present value of the right to receive an annual income of a series of pounds at a given rate of compound interest (i) for a given number of years (n).

As the expanded name — *The Present Value of £1 pa* — suggests, the YP formula is actually a series of present value calculations applied to a set of individual cash flows of £1 pa and the formula, therefore, is in effect a summation of a series of present value calculations. The YP, or PVPA, or PVpa formula is:

$$YP = \frac{1 - PV}{i}$$

So, if you wanted to know the capital value of an income flow of £10,000 pa over four years at an interest rate of 8% this could be found as follows:

$PV = 1/(1+i)^n$
So, $PV = 1/(1+0.08)^4$
$PV = 0.73503$
Therefore, $YP = (1 - 0.73503)/0.08$
$YP = 3.3121$

This means that an investment producing £1 pa each year for four years at a rate of interest of 8% is worth £3.3121. So, £10,000 a year is be worth £33,121.

Proof:

Year	Income		PV		
1	£10,000	×	0.9259	=	£9,259
2	£10,000	×	0.8573	=	£8,573
3	£10,000	×	0.7938	=	£7,938
4	£10,000	×	0.7350	=	£7,350
Capital Value					£33,120

This demonstrates that the YP figure is the sum of a series of individual present value of £1 calculations.

The Annuity £1 will purchase

The *Annuity £1 will purchase* is the opposite of YP. The Annuity of £1 is the income that a single investment of £1 will purchase at a given rate of compound interest (i) for a given number of years (n). The formula for The Annuity of £1 (or Annual Equivalent of £1) is the reciprocal of the YP formula. So, if YP = (1 – PV)/i, the Annuity formula is:

$$\text{Annuity} = \frac{i}{1 - PV}$$

Annuities are often purchased as investments by people who, on retirement, wish to convert a lump sum capital payment to an income flow. For example, what is the annual income (annuity) receivable for 32 years, which a single investment of £1 will purchase, if interest is required on capital at 5% pa.

Annuity = i/(1 – PV)
Where PV = 1/A
And A = $(1 + i)^n$
So, A = $(1 + 0.05)^{32}$
A = 4.7649
PV = 1/A = 1/4.7649
PV = 0.2099

Substituting the PV value back into the annuity formula:

Annuity = 0.05/(1 – 0.2099)
Annuity = 0.05/0.7901
Annuity = 0.0633

This means that £1 will buy an annuity of just over 6 pence pa.

Seen from the perspective of the lender, a mortgage is actually an annuity; the lender gives a capital now sum to the homeowner which is used for the purchase of a house and, in return, receives regular income payments over a period of time.

For example, a building society proposes to lend £200,000 to someone who hopes to buy a house for £250,000. How much will the house buyer have to pay the building society each year, if the loan is for 25 years and the mortgage rate is 7%?

Mortgage loan (ie Capital now)	£200,000
× Annuity £1 will purchase over 25 years @ 7%	0.0858
Annual mortgage payment (ie income)	£17,160pa

Note

- This type of mortgage is called a repayment mortgage. Other types of mortgage use a different investment to recover the capital at the end of the loan: see glossary in Chapter 2.
- 0.0858 is made up of 0.07 + 0.0158 (ie interest on loan + repayment of loan).
- So, £17,160 is made up of £14,000 + £3,160 (ie interest on loan + repayment of the loan).
- If the Building Society invests £0.0158 at the end of each year for 25 years at 7% it will amount to *future capital* of £1. If it invests £3,160 at the end of each year for 25 years at 7% it will amount to £200,000, enough to pay back the loan.

- £200,000 capital *now* is equivalent to an income of £17,162pa over 25 years at 7%.
- If the interest rate changes during the mortgage term the new payment can be found by calculating the loan outstanding at that time by multiplying the payment by the YP at the old interest rate for the remaining term of the mortgage, then multiplying the result by The Annuity £1 will purchase at the new interest rate for the remaining length of the loan.

Calculators and Valuation Tables

Valuers need to be adept at solving problems using the above formulae. So ability to use a financial or scientific calculator is important. There are many cheap models on the market but the main considerations are that any calculator used for solving valuation formulae must have the following functions:

- Reciprocal ($1/x$ or x^{-1})
- Power (x^y or y^x or Λ)
- Root ($x\sqrt{}$ or $x^{1/y}$ or the inverse of the power function).

The ability to change a figure to a negative or a positive (+/–) is useful, but if it is not available on your calculator you can always multiply by –1 instead.

As an alternative, all of the above calculations can be undertaken using valuation tables, such as *Parry's Valuation and Investment Tables* (see further reading, below). Tables are a valuable tool and many would argue that they are quicker and more accurate to use. However, one issue that the tables do not address is situations where the increments between interest rates are very small, or where a fraction of a year is required, and so the calculator has the benefit of total flexibility. Valuers should probably be adept at using both methods of calculation.

The following sample tables illustrate the functionality of valuation tables. They are calculated using a spreadsheet and the appropriate formulae.

Years	Amount of £1 @		Present Value of £1 @	
	3.50%	10.00%	3.50%	10.00%
	£	£	£	£
1	1.0350	1.1000	0.9662	0.9091
5	1.1877	1.6105	0.8420	0.6209
10	1.4106	2.5937	0.7089	0.3855
20	1.9898	6.7275	0.5026	0.1486
30	2.8068	17.4494	0.3563	0.0573
40	3.9593	45.2593	0.2526	0.0221
50	5.5849	117.3909	0.1791	0.0085
100	31.1914	13780.6123	0.0321	0.0001

Notice:

- how sensitive the figures are to the rate of compound interest
- how small the PV figures are after 50 years.

Years	Amount of £1 pa @		Sinking Fund @	
	3.50%	10%	3.50%	10%
	£	£	£	£
1	1.0000	1.0000	1.0000	1.0000
5	5.3625	6.1051	0.1865	0.1638
10	11.7314	15.9374	0.0852	0.0627
20	28.2797	57.2750	0.0354	0.0175
30	51.6227	164.4940	0.0194	0.0061
40	84.5503	442.5926	0.0118	0.0023
50	130.9979	1163.9085	0.0076	0.0009
100	862.6117	137796.1234	0.0012	0.0000

Notice:

- the figures are sensitive to the rate of compound interest
- APA figures grow at an accelerating rate
- the Sinking Fund figures are very small.

Years	Years Purchase @		Annuity £1 will purchase @	
	3.50%	10%	3.50%	10%
	£	£	£	£
1	0.9662	0.9091	1.0350	1.1000
5	4.5151	3.7908	0.2215	0.2638
10	8.3166	6.1446	0.1202	0.1627
20	14.2124	8.5136	0.0704	0.1175
30	18.3920	9.4269	0.0544	0.1061
40	21.3551	9.7791	0.0468	0.1023
50	23.4556	9.9148	0.0426	0.1009
100	27.6554	9.9993	0.0362	0.1000
Perpetuity	28.5714	10.0000	0.0350	0.1000

Notice:

- The figures are sensitive to the rate of compound interest
- YP figures grow quickly but then reach a ceiling figure
- The Annuity figures are very small.

Years Purchase in perpetuity

One final formula used to convert income flows into sums of *capital now* is the Years Purchase in perpetuity (abbreviated to YP in perp). We will see in Chapter 14 how this formula is used to value the perpetual income flows which typically occur in freehold interests. We have seen that the YP, or PVpa, for a given number of years is represented by the formula:

$$YP = \frac{1 - PV}{i}$$

So, the YP for 20 years at 10% can be found by taking the PV value from the above table to be

(1 – 0.1486)/i
and so the YP is (1 – 0.1486)/0.1 = 8.514

The YP for 100 years at 10% can be found by taking the PV value from the above table to be (1 – 0.0001)/i
and so the YP is (1- 0.0001)/0.1 = 9.999

In other words for longer periods of time the PV value gets progressively smaller so that, for perpetual income flows (in practice anything above about 100 years), it will be zero. The YP formula therefore can be rewritten as:

$$\text{YP in perp} = \frac{1}{i}$$

In other words, for a perpetual income flow the YP is the reciprocal of the interest rate.

Different YPs in perpetuity are shown below.

Interest Rate	YP in Perp	Interest Rate	YP in Perp
20%	5.00	10%	10.00
19%	5.26	9%	11.11
18%	5.56	8%	12.50
17%	5.88	7%	14.29
16%	6.25	6%	16.67
15%	6.67	5%	20.00
14%	7.14	4%	25.00
13%	7.69	3%	33.33
12%	8.33	2%	50.00
11%	9.09	1%	100.00

Notice:

- the higher the interest rate (or yield) the lower the YP
- there is an inverse relationship between the YP and the interest rate.

Further reading

Parry's Valuation and Investment Tables, Davidson A W, 12th ed, EG Books, 2000

The Income Approach to Property Valuation, Baum A, Nunnington N, Mackmin D, 5th ed, EG Books, 2006 particularly Chapter 1.

14 Freehold Investment Valuations

This chapter introduces the serious business of valuing freehold investments. In looking at the time value of money in Chapter 13 we introduced the formula for converting income flows into capital now. This is referred to as the Present Value of £1 pa, or PVpa for short. This formula, which is used for the capitalisation of property incomes or rents, is more usually called the Years Purchase or YP because the multiplier is the number of years of income it would take to purchase the investment at its current value.

A freehold interest is the highest form of ownership. A freeholder can use a property to generate investment income by letting the property to a tenant, thus conferring rights of occupation for a fixed and certain period of time.

There are three distinct types of freehold investment: the fixed perpetual income, the fully let freehold and the reversionary freehold. Valuing all three involves using the YP formula adopting a given discount rate. Identifying the appropriate discount rate is one of the most challenging tasks the investment valuer has to undertake because this rate (also referred to by valuers as the yield) has to reflect all the risks attached to a particular property, its tenant and its location. It also has to reflect the growth potential in the income. For this reason it is referred to as the 'all risks yield' and is usually derived from analysis of market transactions.

The reversionary freehold is arguably the most problematic of the three freehold investment types. This is partly because market evidence based on identical properties is unlikely to exist and this requires the valuer to exercise considerable judgment. There are three common variations of methods of valuing reversionary freeholds, the Term and Reversion, Hardcore and Equivalent Yield approaches. Each of these methods will be illustrated in this chapter.

When faced with an investment valuation it is always a good idea to draw a diagram showing the income flow over time. It is normal to plot time in years on the horizontal axis and income in £pa on the vertical axis. The point of intersection of the two axes represents the valuation date. In most case this will be the present time. The basic diagram is shown in Figure 14.1.

Figure14.1 The basic time diagram

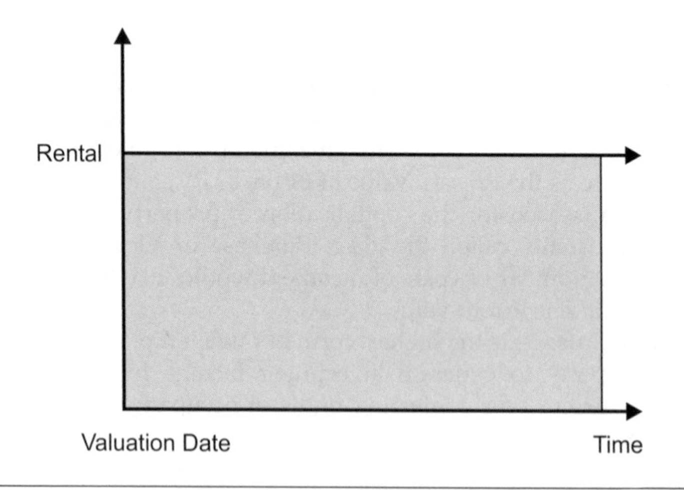

Example 14.1

The fixed perpetual income

Value the freehold interest in a site leased to a developer for an unexpired term of 125 years at a ground rent of £10,000pa, net of outgoings. The rent is fixed for the duration of the lease. An office block was erected by the developer on the site and is let to a tenant at £50,000pa.

The first step is to draw a time line diagram:

Figure 14.2 Fixed perpetual income

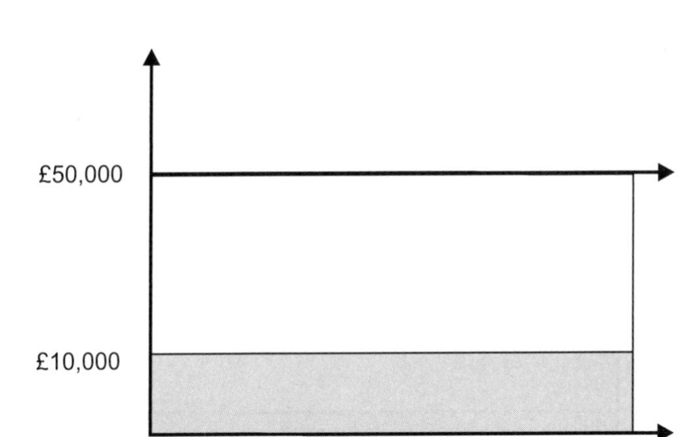

The shaded area represents the freehold interest. This can be seen to be an income of £10,000pa which will last for 125 years. For now we don't have to worry about what happens at the end of the initial lease because this is so far away that it can be regarded as a perpetual income. Note that the rent of £50,000pa is received by the head-leaseholder developer, not the freeholder.

Valuation of freehold interest

Net income	£10,000pa
× YP in perpetuity ('perp' for short) at 12%	8.3333
Market Value of freehold interest	£83,333
say	£83,000

As this is effectively a perpetual income it is appropriate to capitalise this by using the years purchase in perpetuity, which is the inverse of the yield (1/i).

The yield is taken at 12% and this is based on an analysis of open market transactions. This is relatively high because a long fixed income will be highly inflation prone. (This is rather artificial because long leases without rent reviews are very unusual in the modern commercial property market, although examples granted in the past will be encountered on occasion, and houses and flats are commonly let on long leases at fixed nominal ground rents).

The fully let freehold

Example 14.2

Value the freehold interest in a modern office block recently let at its market rent on full repairing and insuring terms of £100,000pa to a government department. The lease contains a clause to review the rent every five years.

The income flow is illustrated in Figure 14.3.

Figure 14.3 The fully let freehold

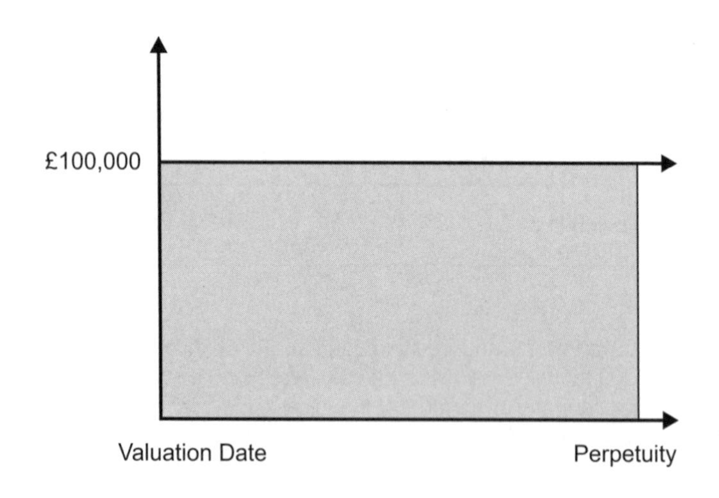

This, like all freeholds, is a genuinely perpetual income flow even though the diagram has to stop somewhere! The income of £100,000pa will be valued using the Years Purchase in perpetuity which is the inverse of the yield (1/i). All the valuer has to do is find the yield and, in ideal circumstances, this will be found using the analysis from open market transactions.

Assuming this yield is 5%, the valuation would be as follows:

Valuation of fully let freehold

Market rent	£100,000pa
× YP in perp at 5%	20
Market Value of freehold interest	£2,000,000

This is quite an attractive investment and so the yield is relatively low. For one thing the lease has regular rent reviews and this means that the freeholder will be able to benefit from any rise in value during the lease and secondly it is let to a government department which would be regarded as a good covenant. This simply means that the tenant is of good standing and is very unlikely to default and so the income under the lease is very safe (or secure).

There are a number of very important assumptions wrapped up in this apparently simple valuation.

- The lease will be for a term of years but the assumption is that at the end of the lease term the property will be re-let, either to the existing tenant or to a new tenant, at the market rent. Assuming this happens at the end of every lease, the income will, in theory continue into perpetuity. In practice because of the effect of the time value of money (the longer you have to wait for an income the less it will be worth in today's terms) for most purposes any income lasting for 99 years or more is regarded as perpetual.
- The income axis does not show any change in rental income and yet we know there will be rent reviews every five years and normally we expect the rent to increase when these occur as the rent paid by the tenant is revised to the then market rent. This is called the fixed income assumption. Because it is difficult to predict in advance how much we can expect the rent to increase, valuers seem to prefer to ignore these unknown future changes, or at least not to make them explicit.
- Future increases in rent however are, in fact, taken into account within the yield and the general rule is that the higher the expectation of future rental growth, the lower will be the yield. Remember that the YP in perpetuity is the inverse of the yield so a low yield will produce a high YP multiplier which in turn will produce a higher market value.

In relation to this last point, it is perhaps helpful to understand that many of these valuation methods were developed during the early part of the 20th century, a time when rental growth was all but unknown. In those days and even through to the 1950s and 60s it was not unusual for property to be let on leases of up to 42 years with a single review at 21 years or even no review at all. In these circumstances the fixed income assumption was an accurate reflection of the anticipated future income flow and this made sense. When, in the post war era, rental growth became the norm, rather than change the valuation methodology, valuers chose to stick with the fixed income assumption, allowing for growth by making adjustments to the yield. These methods of valuation are sometimes referred to as traditional or conventional or growth implicit.

What about inflation and growth?

This issue is illustrated in Figure 14.4. Inflation impacts on the purchasing power of income so that a fixed income investment is relatively unattractive at times of high inflation. Inflation is a downward pressure on the real value of an income flow.

In recent years we have grown used to inflation at relatively low levels but it is worth remembering that there have been times in the not too distant past where inflation reached well into double figures. If inflation is running at 10% pa purchasing power will effectively halve every seven years or so (the Amount of £1 in seven years at 10% is approximately 2, and the PV of £1 in seven years at 10% is approximately 0.5).

Figure 14.4 Real income flows

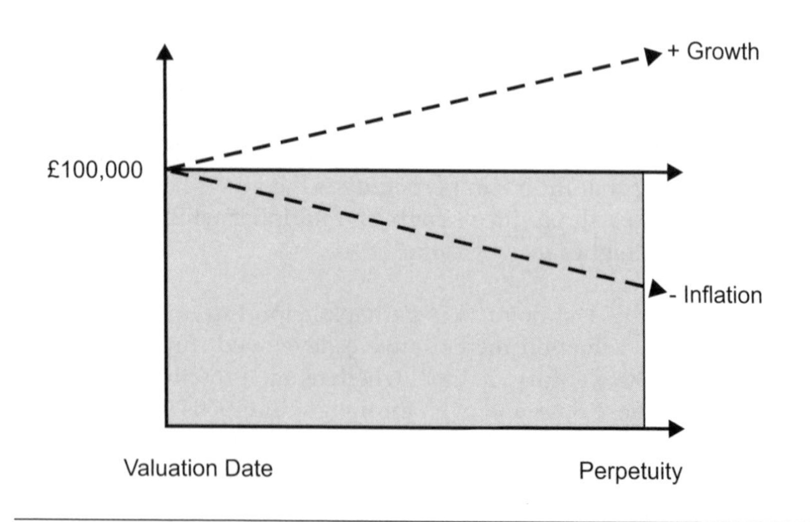

However this is countered by the tendency of property rents to rise over time. This is usually referred to as growth. In fact the tendency for property incomes to demonstrate growth is one of the key attractions of property as an investment.

Growth arises because of the balance between supply and demand. So long as the general level of economic activity is growing then property values will tend to follow suit over the longer term.

The impact of this is shown in Figure 14.5. If growth exceeds inflation over the medium to long term, then the real value of property income flows will tend to increase. This increase will be captured by regular rent reviews which recalibrate the income in terms of the then current market rent. Strictly speaking the income flow between rent reviews will be eroded by inflation but this is normally more than balanced by the level of growth over the long term.

So the income generated by such a property investment is said to be proof against inflation or inflation proof. Conversely a property investment let on a long lease with no rent reviews to capture growth in rents over time would be said to be prone to inflation or inflation prone. Inflation proof investments tend to be seen by the market as more attractive than inflation prone investments and immunity from or exposure to inflation risk is one of the key factors in determining the all risks rate or discount rate or yield.

Figure 14.5 What if growth exceeds inflation?

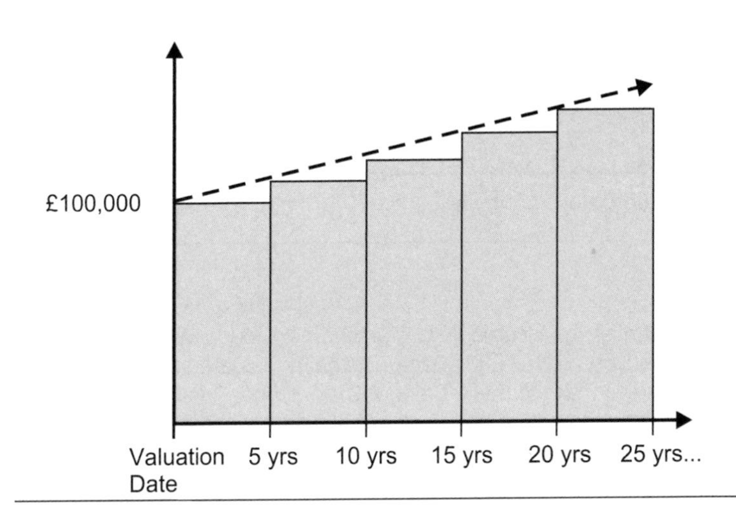

Example 14.2 above looked at a property investment which was let at the market rent. This may be because the valuation date coincides with a periodic rent review or a new letting. However it is far more likely that the valuer will encounter a property at some point in time between rent reviews. This is known as the reversionary freehold and is illustrated in Figure 14.6.

Example 14.6

The reversionary freehold

Value the freehold interest in a modern office block. One year ago the office was let to a government department at £80,000pa. The lease contains a clause to review the rent every five years. The market rent is £100,000pa.

Figure 14.6 The reversionary freehold

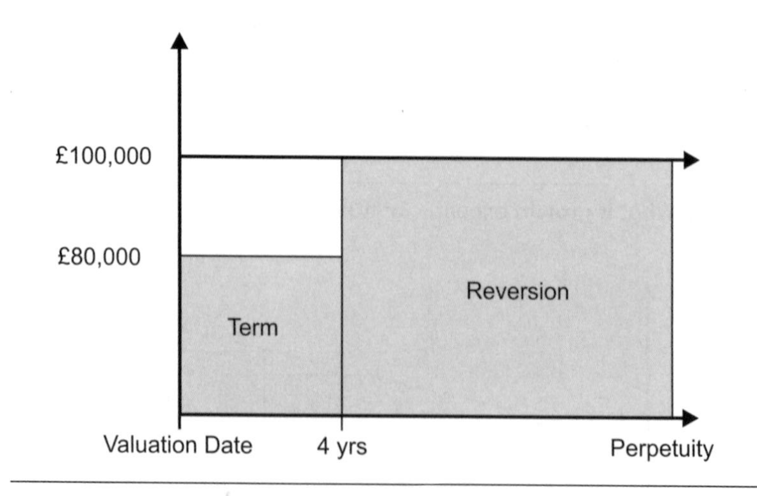

Figure 14.6 illustrates the income flow. Over the next four years, until the next review, the rent will be fixed at £80,000pa. Thereafter, it is assumed that the rent will increase (or revert) to the market rent of £100,000pa. The first four years is referred to as the term and the period after the review (when the rent reverts to the market rent) is known as the reversion. Not surprisingly this method of valuation is usually referred to as the term and reversion approach.

Valuation of freehold interest — term and reversion method

Term			
Rent passing		£80,000pa	
× YP 4 years at 4.75%		3.5666	
			£285,328
Reversion			
Market rent		£100,000pa	
YP in perp at 5%	20		
× PV in 4 years at 5%	0.8227		
× YP in perp deferred 4 years at 5%		16.4540	
			£1,645,400
Market Value			£1,930,728
say			£1,930,000

There are a number of points to note about this valuation.

- The term income is thought to be more secure than average. In other words a tenant paying £80,000pa for a property worth £100,000pa is less likely to default on rental payments because they are enjoying what is, in effect a profit over the next four years. Valuers will sometimes take this into account by adjusting the all risks rate downwards slightly (from 5% to 4.75% in the example).
- The reversion is valued at the all risks rate of 5% because it is the same income flow as the fully let freehold (see Example 14.2). In times of recession the yield for the reversion may be adjusted upwards a little because the investment is slightly more risky as the tenant may leave at the end of the lease and there could be a void period before a new tenant rents the property (although this factor may already be reflected in the all risks yield, so care must be exercised).
- Although there is an expectation of growth over the next four years, it is assumed that the reversionary rent will be the current market rent. This is because the traditional investment method works on a current value current cost assumption. As noted earlier, increases in rental value are taken account of in the all risks yield. The higher the potential level of anticipated growth the lower the yield.
- The value of the reversionary part of the investment will not be realised for another four years so this tranche (or portion) of value is deferred during this waiting period by applying the Present Value of £1 formula to bring the capital value back to the valuation date. As an alternative to multiplying the market rent by the YP and the PV for the length of the term the years purchase in perp deferred figure can be found in valuation tables, or by multiplying the YP in perpetuity by the PV of £1 for the length of the term:

$$\text{YP in perp deferred} = \frac{1}{i} \times \frac{1}{(1 + 1)^n}$$

Where:
i = the yield
n = the length of the term.

Figure 14.7 explores a slightly different way of separating out the different income flows. Here, rather than the vertical separation of income flows of the term and reversion method the income flows are divided horizontally into slices. The bottom slice, or hardcore, is the first layer of income. This is regarded as more secure and so is valued using a slightly lower discount rate or yield. The top slice or marginal income is regarded as less secure and is valued at a higher discount rate. This is because if there was to be a short term recession in market values and rents declined by say 10%, the market rent would fall from £100,000pa to £90,000pa. This would potentially eat into the top slice of income but would not impact on the more secure bottom slice.

Figure 14.7 Hardcore or layer method

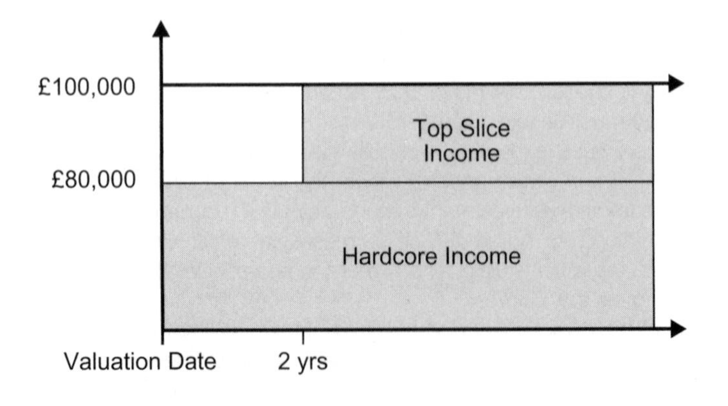

Valuation of freehold interest — hardcore method

Bottom slice			
Rent Passing		£80,000pa	
× YP in perp at 4.75%		21.0526	
			£1,684,208
Top slice			
Market rent		£100,000pa	
less rent passing		£80,000pa	
Top slice Income		£20,000pa	
YP in perp 5%	20.000		
× PV in 4 years at 5%	0.8227		
× YP in perp deferred 4 years at 5%		16.4540	
			£329,080
Market Value			£2,013,288
say			£2,000,000

This method can be useful where property is over rented, in other words where the rent paid by the tenant is more than the market rent (see Example 14.7). This tends to happen during periods of recession and in local areas suffering economic decline. With falling rental values the valuer might want to place a heavy discount against the top slice income (sometimes referred to as the froth income) because of the likelihood of tenant default which will result in the income falling to the market rent. The quality of the tenant's covenant will also influence the choice of discount rate or yield.

This vertical (term and reversion) or horizontal (hardcore) division of income provides valuers with the means of distinguishing between different incomes with different levels of security. Many valuers now take the view that such distinctions are rather artificial and that small adjustments of discount rates are intuitive and cannot be supported by market evidence. Indeed, they may lead to illogical results because slight changes in yield may have a major influence on the resulting valuation figure. This introduces a third possible method of valuation which does not differentiate the different income flows but adopts the same yield throughout. This is known as the equivalent (or same) yield approach.

Valuation of freehold interest — equivalent yield method

Term			
Rent passing		£80,000pa	
× YP 4 years at 5%		3.5460	
			£283,680
Reversion			
Market rent		£100,000pa	
YP in perp at 5%	20.000		
× PV in 4 years at 5%	0.8227		
× YP in perp deferred 4 years at 5%		16.4540	
			£1,645,400
Market Value			£1,929,080
Say			£1,930,000

The equivalent yield method has much to recommend it. It is simple. The valuer has only to work with a single discount rate which is applied throughout the calculation. It also recognises the artificiality of minor and intuitive yield adjustments for which there is no empirical market evidence.

Over rented freehold interests

Example 14.7

Value the freehold interest in a shop on the fringe of a town centre held on a lease with two years unexpired at £10,000pa on full repairing and insuring (FRI) terms. The area has declined and the market rent is now only £8,000pa.

The freehold all risks yield is 9%, reflecting a secondary or tertiary shop in a poor area.

The top slice of income is regarded as at high risk and is therefore discounted at a higher discount rate producing a lower valuation figure.

Figure 14.8 Over rented freehold

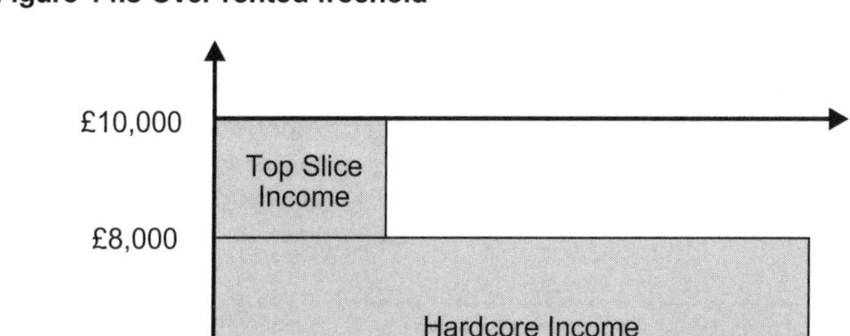

Valuation of over rented shop		
Hardcore Income		
Market rent	£8,000pa	
× YP in perp at 9%	11.1111	
		£88,889
Top Slice Income		
Rent reserved	£10,000pa	
less market rent	£8,000pa	
Top slice income	£2,000pa	
× YP 2 years at say 11%	1.7125	
		£3,425
Market Value		£92,314
say		£92,000

Which to use — term and reversion or equivalent yield?

Whether using the term and reversion or equivalent yield methods, provided that the term is short and the rent passing is close to the market rent, which will normally be the case with modern leases, the different approaches to valuing reversionary freeholds outlined above will all produce virtually the same valuation figure.

Inflation prone income flows

Example 14.8

The penultimate valuation in this chapter on freeholds is included to illustrate the way in which yields can be used to reflect the extent to which income flows are inflation prone. Figure 14.9 is a freehold interest let on a long lease without review. The existing lease has 20 years to run. At the end of this lease it is assumed that the property will be re-let at the market rent on more conventional terms with rent reviews every five years.

Figure 14.9 Freehold let on long lease

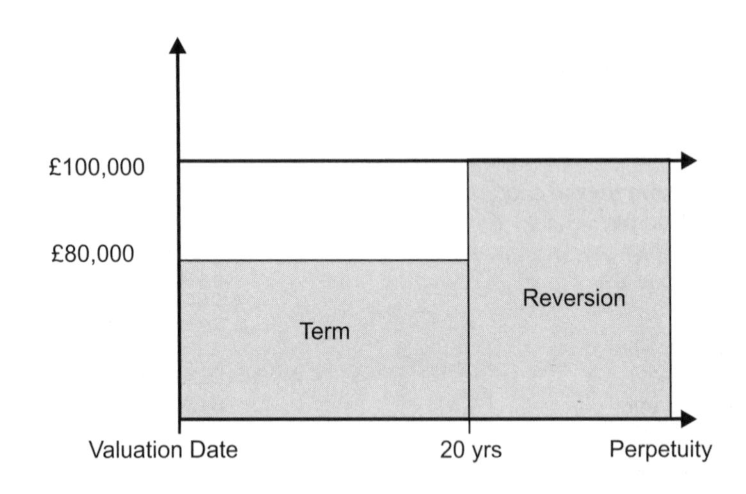

The valuation used is a term and reversion calculation but this time note that the relative inflation proneness of the term income can be reflected by adopting a relatively high discount rate compared to the all risks rate used to value the reversion.

Valuation inflation prone income

Term		
Rent passing	£80,000pa	
× YP 20 years at 7%	10.5940	
		£847,520

Reversion			
Market rent		£100,000pa	
YP in perp at 5%	20.000		
× PV in 20 years at 5%	0.3769		
YP in perp deferred 20 years at 5%		7.5380	
			£753,800
Market Value			£1,601,320
Say			£1,600,000

Reversionary freehold with outgoings

Example 14.9

Finally, where the rent collected is subject to outgoings (see Chapter 7) the rent collected must be reduced to reflect the fact that landlord's liability to carry out repairs and/or insure the property will mean that the income received is not net.

You are required to value the freehold interest in a small shop on a residential estate. The market rent is £5,000pa on FRI terms. The premises are let on a lease with three years unexpired at £4,800pa on internal repairing (IR) terms.

Valuation of freehold where the rent is subject to outgoings

Term			
Rent reserved IR terms		£4,800pa	
less freeholder's outgoings:			
External repairs say 10% market rent	£500pa		
Insurance say 2% market rent	£100pa		
Management say 10% rent reserved	£480pa		
		£1,080pa	
Rent paid on FRI terms		£3,720pa	
× YP in 3 years at 9%		2.5313	
			£9,416
Reversion			
Market rent FRI terms		£5,000pa	
× YP in perp deferred 3 years at 9%		8.5798	
			£42,899
Market Value			£52,315
say			£52,000

Note that

- IR terms means that the tenant pays for internal repairs, leaving the landlord responsible for external repairs and insurance. Management is allowed if there are any other outgoings. Management is usually ignored for FRI rents, arguing that it is allowed for in the all risks yield
- The yield of 9% reflects the secondary/tertiary quality of the investment
- The yield used to value the term could be slightly increased because the liability for outgoings increases the risk to the landlord because the rent is fixed but the outgoings might be more than expected.

Summary

This chapter has considered a range of possible variants of freehold income flows generated by conventional property investments. In all cases the method consists of a number of steps.

- Identify the income flows and show these over time on a diagram.
- Where there are varying income flows decide how to separate these and apply appropriate discount rates to find the present value of each income flow.
- Differentiation between discount rates will depend upon the relative security or risk attached to that income flow.
- One of the key risks will be the *inflation proneness* of the income flow or its *inflation proofness*. An inflation prone income will decline in real value terms over time whereas an inflation proof income flow will keep pace with and even exceed inflation.
- Use relatively high discount rates for inflation prone income flows and relatively low rates for inflation proof income flows.
- Where high levels of rental growth are anticipated relatively low rates are adopted.

Further reading

The Income Approach to Property Valuation, Baum A, Mackmin D and Nunnington N, 5th ed, EG Books, 2006

15 Conventional Leasehold Investment Valuations

The valuation of leasehold interests is quite complicated technically, and raises a number of issues about valuation technique. Several variations of the traditional investment method of valuation are in use, and some of these have been criticised by academics over a considerable number of years. This chapter discusses the options available but does not come to any definitive conclusion about the appropriate technique to use. The further reading section lists a paper which gives the view of a leading academic.

A lease or tenancy gives the tenant a package of rights and responsibilities in relation to the subject property, and this package of rights is what the surveyor is valuing.

Factors affecting the value of leasehold interests

The following questions have to be answered before a valuation of a leasehold interest can be prepared.

1. What is the start date and the length of the lease? The commencement date will often be different from the date the lease was granted. In older leases look out for phrases like "to hold the same for a term of ... years from ...". This information is used to find the unexpired term of the lease, that is how long the tenant has left from the valuation date to the end of the lease.
2. Is there a break clause allowing either party to end the lease early?

3. How much rent is payable and what is the timing of payments? The lease will specify whether the rent is paid in advance or arrears and the payment period. In practice for valuation purposes it is usual to assume that rent is payable annually in arrears.
4. Is there a rent review clause? If so, when can the rent be reviewed, how is the new rent to be assessed, and is the clause upwards only or can the rent fall on review?
5. Who pays for repairs, insurance, rates and other taxes? Are any of the landlord's costs recovered from the tenant by a service charge?
6. Is there any restriction on assignment (transferring the lease) or sub-letting (granting a sub-lease for a term less than the head tenant's lease, possibly of only part of the property)? This is particularly important if the covenant is absolute. If the covenant requires the landlord's consent before assignment or sub-letting, section 19 of the Landlord and Tenant Act 1927 provides that consent shall not be unreasonably withheld. Some leases require a tenant who wishes to assign to surrender the lease to the landlord.
7. How much rent is collected from a sub-tenant or, if the property is vacant or occupied by the tenant, what rent could be collected if the property was to be let? The rent collectable takes into account the covenants in the lease, particularly the user clause which may prevent the property being used for the purpose which would give its highest and best value.

The market for the sale of non-residential leasehold interests is significantly smaller than that for freeholds, with far fewer transactions. There are therefore often few if any comparables, and leasehold interests are normally valued by reference to the freehold interest using the investment method.

Figure 15.1 shows how the freehold and leasehold interests fit together on the time line diagram.

Although the investment method is used to prepare valuations, most non-residential leasehold interests are not bought as investments. The purchasers are either occupiers, who want to use the property in connection with a business, or the freeholder, who wishes to merge the freehold and leasehold interests so that the property can be redeveloped, sold with vacant possession or re-let on modern lease terms. This is particularly true of short leases, where the costs of acquisition are high in relation to the value of the interest, making it difficult for an investor to get a satisfactory return.

Figure 15.1 Leasehold interests

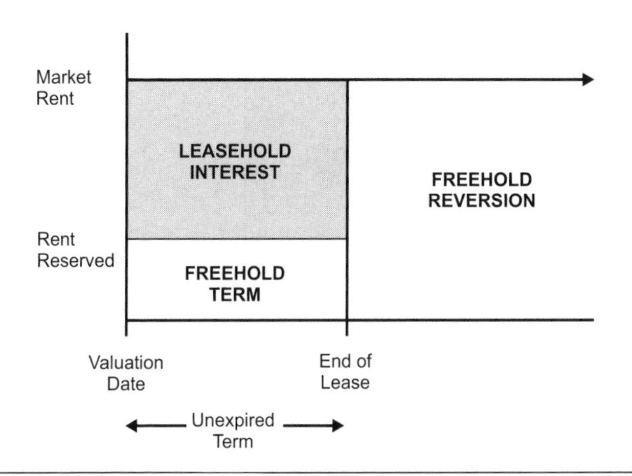

The basis of the valuation

The traditional method for valuing a leasehold interest is to capitalise the profit rent. The profit rent is the difference between the market rent if the property is vacant or occupied by the tenant, or the rent collected if the property is sub-let, and the rent paid by the tenant on the same terms.

A profit rent may arise for any or all of the following reasons.

1. Time passing since the lease was granted. The market rent of property usually increases as time goes on, and so will normally be more than the rent the tenant pays. In a recession it is possible for the market rent to fall below the rent reserved, particularly if there is an upwards only rent review. In these circumstances the property is said to be over rented, the profit rent will be negative and the leasehold interest is a liability with a negative value.
2. The tenant paid a premium (see glossary in Chapter 2).
3. The tenant has carried out improvements, for example by fitting out the shell of a new shop, or by modernising the premises. Whether the value of any works carried out by the tenant is included in the rent payable on a rent review depends on the terms

Figure 15.2 Leasehold profit rent

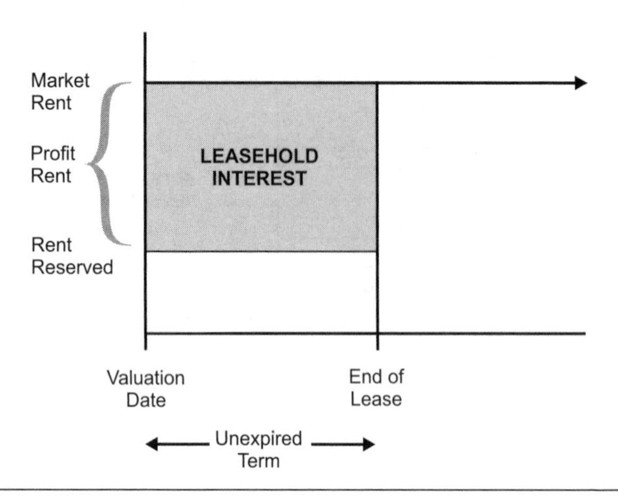

of the lease. Modern leases normally provide for non-contractual improvements by the tenant to be disregarded (otherwise the tenant pays twice, once to carry out the work and then a second time by having the rent put up on the next review to reflect the increased value of the property). However, at common law there is a presumption that, in the absence of an express provision, the rent on review is assessed as the premises stand.

4. The letting was not at arm's length — the tenant might be a member of the landlord's family, or part of the same group of companies, or there could be some other reason why the landlord granted a lease at a low, concessionary, rent, for example local authorities sometimes make concessionary lettings to groups who will provide a service to the local community.

Estimating the profit rent

The profit rent is found by taking the market rent (or the rent collected if the property is sub-let) and deducting the rent paid on the same terms.

Example 15.1

Example 14.6 was a valuation of the freehold interest in a modern office block. One year ago the office was let to a government department at £80,000pa on FRI terms. The lease contains a clause to review the rent every 5 years. The market rent is £100,000pa on FRI terms. What is the leaseholder's profit rent?

Market rent, on FRI terms	£100,000pa
less rent paid, on the same terms	£80,000pa
Profit Rent (or net income to the leaseholder)	£20,000pa

It is important that the market rent and the rent paid are on the same terms. If they are not an adjustment must be made for outgoings (see Chapter 7).

Example 15.2

You are required to value the leasehold interest in a small shop on a residential estate. The market rent is £5,000pa on FRI terms. The premises are let on a lease with three years unexpired at £4,800pa on internal repairing terms. What is the leaseholder's profit rent?

To find the profit rent, the rent paid and the market rent must be on the same terms. The simplest way to perform the calculation is to adjust the rents so that they are both net. It is possible to go the other way and add outgoings to the FRI rent, but this needs careful thought, can be confusing and may lead to mistakes.

Market rent FRI			£5,000pa
Rent paid IR terms		£4,800pa	
less outgoings:			
External repairs, say 10% market rent	£500pa		
Insurance, say 2% market rent	£100pa		
Management, say 10% rent reserved	£480pa		
		£1,080pa	
less rent paid on FRI terms			£3,720pa
Profit Rent			£1,280pa

Having found the profit rent it only remains to capitalise it, that is convert it into the market value. The problem is the yield and the type of YP to use.

Deciding on the length of the leasehold term

In simple cases the rent paid will be fixed for the whole of the lease. The tenant will therefore enjoy the profit rent for the whole of the unexpired term.

If a landlord grants anything other than a short lease (say three to five years) the value of the rent collected will be eroded by inflation. It is therefore normal to include a rent review clause in leases which allows the rent to be increased, commonly to the market rent at the date of review. Rent review periods are often every five years for prime property; a three year rent review pattern may be found for poorer investments.

Because the rent paid will increase to the market rent on review the tenant will cease to have a profit rent, and the value of the leasehold interest in investment terms is nil or nominal at that stage. In these circumstances the unexpired term is taken as the time from the valuation date to the date of the next rent review, even though the tenant will still have an interest in the property until the end of the lease.

If there is a break clause allowing the landlord to end the lease early it is usual to assume that the landlord will exercise the break clause, and the unexpired term is treated as the time to the break clause, or the next rent review to market rent, whichever is earlier.

It should only be assumed that a break clause which can be exercised by the tenant would be exercised if it would be to the tenant's advantage to do so, for example if there is an upwards only rent review which would mean that the rent will be more than the market rent.

The relationship between leasehold and freehold yields

The value of a leasehold interest will diminish as the end of the term approaches. Leases are terminating interests — at the end of the lease the tenant's rights in the premises are over and the leaseholder no longer has an interest in the property. There are a few exceptions to this where statute gives tenants security of tenure, although the tenant will generally have to pay market rent, and so will not have a profit rent.

A leaseholder is less secure than a freeholder — the landlord has rights and can exercise some control over the property, for example by refusing consent to assign, to alter the premises, or to change the use of the property, and in extremis can forfeit the lease if the tenant is in breach of covenant. In addition, the profit rent is top slice income, less secure than the hardcore bottom slice of the freehold (see Chapter 14).

Leasehold interests are, therefore, less attractive to investors and occupiers than freehold interests, and yields on leaseholds can be expected to be higher than for similar freehold property. This reflects the lower market value of leasehold incomes compared to freehold incomes.

Traditionally the yield on a leasehold interest is taken to be 0.5% to 2% above the all risks yield of similar freehold property. The lower figure would be used for a prime property on a long lease, the higher figure is appropriate for a poor property on a fag end lease (a lease that has almost expired).

Adjusting the freehold yield in order to arrive at a yield to value leasehold interests is questionable. Comparison relies on comparing like with like, and while the property involved is the same there are significant differences between freehold and leasehold tenure. Ideally the leasehold yield can be found by comparison with the sale of similar leasehold property. In the absence of leasehold comparables it is suggested that leaseholds can be valued at a yield found by taking the return on gilts (government stock) with a redemption date at the same time as the profit rent from the lease ends, and then adding an appropriate amount for the additional risks involved in investing in leasehold property.

Single Rate, Dual Rate (DR) or Dual Rate tax adjusted (DRT) YPs

Because leases are terminating interests they are valued using a Years Purchase for the number of years that the profit rent lasts, although the YP in perpetuity may be used if the unexpired term is over 99 years. The capital value (market value) on the purchase of the lease can either be seen as being repaid during the term of the lease, in the same way that a lender recovers the capital lent on a repayment mortgage or being recovered through the annual sinking fund element of the YP (see below).

Chapter 13 explained that the formula for the YP was:

$$\frac{1 - PV}{i}$$

Where

PV = the Present Value of £1 = $\frac{1}{(1 + i)^n}$

i = the rate of interest
n = the number of years.

An alternative formula is:

$$YP = \frac{1}{i + SF}$$

Where

SF = Annual Sinking Fund = $\frac{1}{A - 1}$

The alternative formula for YP shows that the YP for a number of years has two elements. The interest rate i gives a return *on* the capital invested, while the sinking fund gives a return *of* the capital invested so that when the lease ends the investor is able to replace the capital invested, see Figure 15.3 below:

Figure 15.3 Return on capital and return of capital

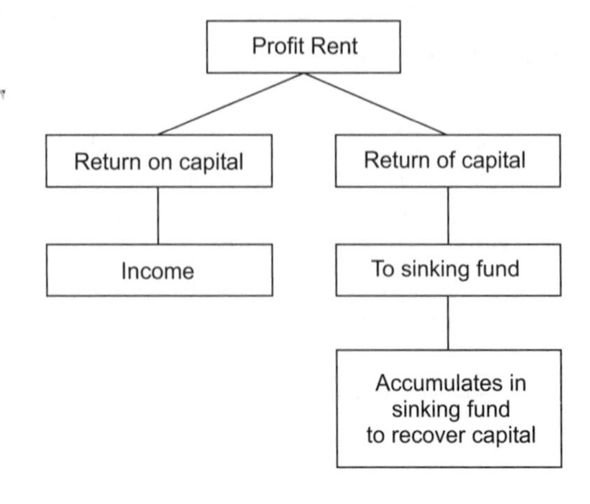

Several different types of YP are currently in use for the valuation of leasehold interests, and it is probably simplest to demonstrate them in the order that they were developed.

Single Rate YP

Single rate YP tables can either be seen as recovering the capital during the term of the lease, or as replacing the capital with a sinking fund at the same rate as the yield on the property, effectively reinvesting in the same property or in a portfolio of similar properties.

Example 15.3

Value the leasehold interest in a modern office block. One year ago the office was let to a government department at £80,000pa. The lease contains a clause to review the rent every 5 years. The market rent is £100,000pa. The freehold all risks rate is 5% (see Example 14.6 in Chapter 14 for the freehold valuation).

For the purposes of this example the leasehold yield will be taken at 6% (1% above the freehold all risks rate).

Valuation of leasehold office block		
Market rent FRI terms	£100,000pa	
less rent paid same terms	£80,000pa	
Profit Rent	£20,000pa	
× YP 4 years at 6%	3.4651	
Value of leasehold interest		£69,302
say		£69,000

Because the leasehold interest is a wasting asset it is worth nothing in four years time, so an investor needs to consider how to replace the capital. There are two approaches.

1. *Instalment approach*

The profit rent of £20,000pa will be used to give a 6% return on the capital outstanding each year, after deducing the return on capital the balance is available to reduce the capital outstanding, so year by year less money is required as a return.

Year	Capital Outstanding	Return on Capital @ 6%	Return of Capital*
1	£69,302	£4,158	£15,842
2	£53,460	£3,208	£16,792
3	£36,668	£2,200	£17,800
4	£18,868	£1,132	£18,868
5	£0		

* profit rent less the return on capital each year (eg £20,000 – £4,158 = £15,842).

So, the capital (£69,302) is returned to the investor during the lease and a 6% return has been made on the money invested.

2. Sinking fund approach

Income from the property		£20,000pa	
less return on capital	£69,302		
@ 6%	× 0.06		
		£4,158pa	
Sum available to invest in sinking fund		£15,842pa	
We can find how much this amount invested each year will accumulate to over a given number of years using the Amount of £1pa			
× APA in 4 years at 6%		4.3746	
Capital recovered			£69,302

So, the sinking fund recovers the capital paid to purchase the leasehold interest at the end of the lease.

Dual Rate YP

The investor or surveyor who takes the instalment approach will accept that the capital has been recovered during the term. The problem with the single rate YP for a surveyor who takes the sinking fund approach is that the sinking fund element of the YP is being notionally reinvested in the leasehold property, which at the end of the lease is not worth anything.

The idea behind dual rate YP tables is that the return of capital should be through some investment other than the property itself. A sinking fund could be set up through an insurance company where the annual payments will be invested, and these will accumulate to return the capital when the lease ends.

The rate of interest used for the return *on* capital is the called the *remunerative rate,* while the sinking fund for the return of capital is invested at a lower, safe, net of tax, *accumulative rate.* Common accumulative rates are 2.5%, 3% and 4%.

The reasons for the use of a historically low rate are:

- tax will have to be paid on the interest earned in the sinking fund. The rates used are *net* of tax. For example, when a 5% gross of tax rate is adjusted for tax at 40% the net of tax rate of accumulation falls to 3%
- a safe average rate of return is required over a long period
- a guaranteed lump sum is required at the end of the lease
- there will be administration costs involved in dealing with a series of small payments
- the sinking fund will probably accumulate to *more* than is required (because interest rates available over the unexpired term may turn out to exceed the low accumulative rate adopted) giving some degree of inflation hedge and making the investment more comparable with a freehold.

Dual rate YP tables were first published in 1895, and the use of dual rate as an alternative to single rate was discussed in text books from 1908, although their use did not become common until after the Second World War.

The formula for a dual rate YP is:

$$YP = \frac{1}{i + SF}$$

Which looks the same as the single rate alternative formula, but now:

i = the remunerative rate

SF = annual sinking fund at the accumulative rate.

Written out in full the formula becomes:

$$YP = \frac{1}{i + \frac{i'}{(1 + i')^n - 1}}$$

Where:
i = the remunerative rate
i' = the accumulative rate
n = the number of years

So there are two interest rates, the remunerative rate and the accumulative rate, hence 'dual rate YP'.

Example 15.4
Value the leasehold interest in a modern office block. One year ago the office was let to a government department at £80,000pa. The lease contains a clause to review the rent every five years. The market rent is £100,000pa. The freehold all risks rate is 5% (see Example 14.6 in Chapter 14 for the freehold valuation).

For the purposes of this example the leasehold yield will be taken at 6% (a little above the freehold all risks rate), allowing for a sinking fund at 3%.

Valuation of leasehold office block

Market rent FRI terms	£100,000pa	
less rent paid same terms	£80,000pa	
Profit Rent	£20,000pa	
× YP 4 years 6% & 3%	3.3442	
Value of leasehold interest		£66,884
say		£66,500

The lease is worth nothing in four years time, but the purchaser's capital is recovered through the sinking fund:

Income from the property		£20,000pa	
less return on capital	£66,884		
@ 6%	×0.06		
		£4,013pa	
Sum available to invest in sinking fund		£15,987pa	
We can find how much this amount invested each year will accumulate to over a given number of years using the Amount of £1pa at the accumulative rate			
× APA in 4 years at 3%		4.1836	
Capital recovered			£66,884

Notice that the capital value has fallen compared to the result of Example 15.3. A greater proportion of the income is required for the sinking fund at the lower remunerative rate to recover the capital, leaving less income to give a return on capital, and a lower capital value.

Dual Rate Tax Adjusted YP

While the tax paid on the interest accumulating in the sinking fund was allowed for by using a *net* of tax accumulative rate, income tax is also payable on the rent received from property. If this is not taken into account the money set aside to be invested in the sinking fund will be insufficient to replace the capital.

Figure 15.4 Dual rate tax adjusted

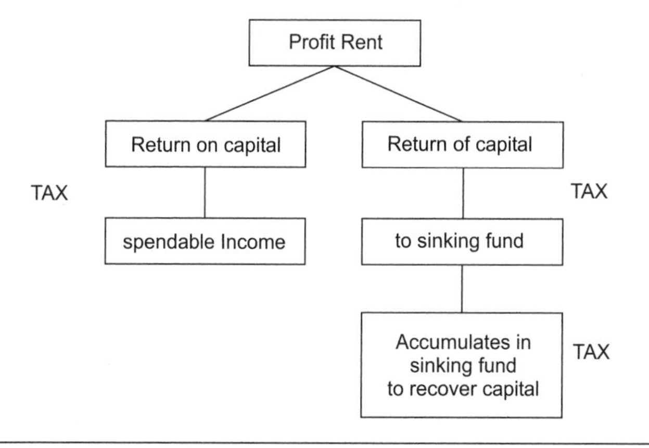

Tax adjustment of dual rate tables was discussed in text books before the Second World War. The first dual rate tax adjusted tables were published in 1949, dual rate was first included in *Parry's Tables* (the then standard set) in the early 1960s, and the technique was in common use in practice by the 1970s.

In Dual Rate Tax adjusted (DRT) tables the sinking fund is multiplied by the tax fraction (Tg) to reflect the need to pay tax on money invested in the sinking fund:

$$Tg = \frac{1}{(1 - t)}$$

Where t is the rate of tax expressed as a decimal.

The formula for a dual rate tax adjusted YP is:

$$YP = \frac{1}{i + SF \frac{1}{(1 - t)}}$$

Where:

i = the remunerative rate
SF = annual sinking fund at the net of tax accumulative rate.
t = the rate of tax as a decimal.

Written out in full the formula becomes:

$$YP = \frac{1}{i + \frac{i'}{(1 + i')^n - 1} \frac{1}{(1 - t)}}$$

Where:

i = the remunerative rate
i' = the accumulative rate, net of tax
n = the number of years
t = the rate of tax as a decimal.

A major issue is deciding what tax rate to use. At the time of writing the basic rate of income tax is 20%, higher rate income tax is charged at 40% and the main rate of corporation tax is 28%. In practice many surveyors have used either 40% (with a 3% sinking fund) or 35% (with a 2.5% sinking fund) irrespective of the tax rates payable at the time.

Example 15.5

Value the leasehold interest in a modern office block. One year ago the office was let to a government department at £80,000pa. The lease contains a clause to review the rent every five years. The market rent is £100,000pa. The freehold all risks rate is 5% (see Example 14.6 in Chapter 14 for the freehold valuation).

For the purposes of this example the leasehold yield will be taken at 6% (a little above the freehold all risks rate), allowing for a sinking fund at 3% adjusted for tax @ 40%.

Valuation of leasehold office block

Market rent FRI	£100,000pa	
less rent paid same terms	£80,000pa	
Profit rent	£20,000pa	
× YP 4 years at 6% and 3% (tax at 40%) (sometimes just written as "6% DRT")	2.1816	
Value of leasehold interest		£43,632
say		£43,500

And the recovery of capital through the sinking fund can be demonstrated as follows:

Income from the property		£20,000pa	
less return on capital	£43,632		
@ 6%	×0.06		
		2,618pa	
sum available to invest in sinking fund gross of tax		£17,382pa	
less tax at 40%		£6,953pa	
sum available to invest in sinking fund net of tax		£10,429pa	
We can find how much an amount invested each year will accumulate to over a given number of years using the Amount of £1pa at the accumulative rate			
× APA 4 years at 3%		4.1836	
Capital recovered			£43,632

Resulting in a significantly lower valuation figure.

Which YP should be used when?

The Valuation of Property Investments, Enever and Isaac, Estates Gazette, 2002, indicates that the majority of valuers favour an accumulative rate of 4% with no tax adjustment. Later Enever and Isaac suggest that the following approaches are appropriate.

Lease length	Valuation method
Less than 5 years	Single rate (or discounted cash flow — see Chapter 16)
5–25 years	Dual rate if there is good demand, otherwise dual rate tax adjusted
25–50 yrs	Dual rate tax adjusted
More than 50 years	Treat as freehold (that is use single rate).

While the dual rate and dual rate tax adjusted approaches appear to have a compelling logic they have attracted considerable criticism from the 1980s onwards.

The combination of two (for dual rate YPs) or three (for dual rate tax adjusted YPs) variables makes analysis of such comparables as exist difficult. Analysing at different sinking fund and tax rates will produce different remunerative rates. The examples above produce three valuation figures for the same leasehold interest:

Single rate	£69,302
Dual rate	£66,884
Dual rate tax adjusted	£43,632

The lower the rate used in the sinking fund and the higher the tax rate the lower the resulting valuation figure.

If a comparable is available it can be analysed to find the remunerative rate provided that the accumulative rate and tax rate are known. Let us assume that the leasehold property in the example has been sold recently for £69,302 (ie 6% single rate).

Adopting a 3% sinking fund with no adjustment for tax		
Income from the property		£20,000pa
less sinking fund to recover capital	£69,302	
× SF 3% 4 years	0.2390	
		£16,565pa
Return on capital		£3,435pa

= £3,435pa/£69,302 = 4.9%, less than the freehold all risks rate, which is not logical!

If the same calculation is carried out adopting a 4% sinking fund with no adjustment for tax the remunerative rate is 5.3%, also below the freehold rate, while if the transaction is analysed using a 3% sinking fund adjusted for tax at 40% the result is a negative return (and an unhappy owner of the leasehold interest) because the entire profit rent isn't enough to recover the capital. This illustrates the difficulty with analysing short leasehold transactions, much of the profit rent is required to replace the capital, and fixing the accumulative rate (and the tax rate for DRT analysis) has a crucial effect on the remunerative rate.

All three approaches are simply trying to explain the YP (which is actually a unit of comparison — see Chapter 9), and the use of dual rate and dual rate tax adjusted tables can be seen as a way of explaining the YP in a way which satisfies the prejudice in favour of trying to have a system which fits in with freehold all risks rates. It may be that using dual rate, or dual rate tax adjusted, tables is simply a way of reflecting the market's distaste for some forms of leasehold interest.

To be an investment there should be a sub-tenant paying rent to the head-tenant investor. The purchasers of short leases are often not investors, they are occupiers. Incoming tenants, particularly of shops, are sometimes prepared to pay an amount over the value of the property calculated by the investment method, this additional payment is called "key money". Potential assignees may be prepared to pay key money because they expect to remain in the premises after the original lease ends, so the cost of obtaining a favourable location can be spread over a period longer than the unexpired term of the lease. In addition the price of a short lease will often be a comparatively small proportion of the cost of opening a business in a new location, and incoming tenants may not think it worth seeking professional advice.

In reality very few, if any, sinking funds are set up. So the dual rate approaches are founded on a fiction. It can be argued that the sinking fund is being used to produce a model illustrating how investors should behave, however purchasers of leasehold interests patently do not behave in this way. If a sinking fund was to be used it will only recover historic cost (there is an argument that the historically artificially low sinking fund rates means that there is some potential for the sinking fund to grow beyond the replacement of the cost of purchase). In any event, what sinking fund rate should be used?

If tax adjusted tables are appropriate, what tax rate should be used? Investors are often tax exempt, for example charities. The rate of tax of an individual potential purchaser may affect the investor's decisions, but the valuer is attempting to find a market value, and so should look at the group of likely purchasers with the lowest tax liability, who will be able to outbid tax payers who pay higher rates of tax.

Negative profit rents

Occasionally the rent reserved will be more than the market rent, or a sub-tenant will pay less than the head rent. The result is a negative profit rent, sometimes called a 'loss rent'. This can occur where an area is declining in value, and is more common in a recession. A negative profit rent is a liability, the market value will be negative. There is no question of seeking to return capital via a sinking fund, a single rate approach should therefore be adopted.

Example 15.6

Value the leasehold interest in a shop on the fringe of a town centre held on a lease with two years unexpired at £10,000pa on FRI terms. The area has declined and the market rent is now only £8,000pa. (See example 14.7 in chapter 14 for the valuation of the freehold interest in these premises.)

The tenant will probably have to pay someone to take it off their hands. In this case a leasehold yield of 10% has been adopted, reflecting a secondary or tertiary shop in a poor area. The yield might be lower, reflecting the cost of borrowing money, which will produce a larger negative value (strange things happen in the world of negative valuations).

Valuation of leasehold interest in shop

Market rent FRI	£8,000pa	
less rent paid same terms	£10,000pa	
Profit Rent	–£2,000pa	
× YP 2 years at 10%	1.7355	
Value of leasehold interest		–£3,471
say		–£3,500

Conclusion

While there are compelling reasons to adopt a single rate approach in the valuation of leasehold interests, and these have been pointed out by academics for at least 20 years, the dual rate and dual rate tax adjusted approaches remain in common use in the UK. The valuation profession is often slow to respond to changes in technique.

The main argument in favour of the dual rate approaches is that, for all their problems, they reflect the market. It may be that this is simply because many valuers continue to use them.

One factor which none of the methods considered above can deal with satisfactorily is the way in which leasehold interests can show complex growth patterns, and this is discussed in Chapter 17.

Further reading

The Income Approach to Property Valuation, Baum A, Mackmin D and Nunnington N, 5th ed, EG Books, 2006, particularly Chapters 2 and 7.

"Dual Rate is Defunct? A review of dual rate valuation, its history and its irrelevance in today's UK leasehold market", Mackmin D, *Journal of Property Investment Finance* 26,1, 2007.

16 Discounted Cash Flow

The discounted cash flow (DCF) technique began as a decision making tool developed by accountants and economists. The system is very closely related to the traditional investment method of valuation which has been developed and used by valuers from the early 17th century onwards.

DCF, in its basic form, is a more simple concept than its sister valuation approach. This simplicity allows the system to be modified easily to allow the introduction of elements into a valuation which the traditional approach does not deal with explicitly (such as inflation), and reduces the complexity of thought required to produce calculations dealing with the effect of taxation and peculiar income flows.

Discounting and cash flows

This is probably a good point to recap on the ideas discussed in Chapter 13, dealing with the concepts of compounding and discounting. As the name suggests DCF is all about discounting future cash flows.

To prepare a DCF the receipts (or benefits) and costs (or liabilities) of the scheme or investment being considered are tabulated, and then discounted.

For example, a development site is available for £1,000,000. Carrying out the development will cost £700,000 at the end of the first year, £500,000 at the end of the second year, and £400,000 at the end of the third year. The completed development will be sold for £3,500,000 at the end of year 4. The table below shows these cash flows discounted at 10%. The aggregate result is called the Net Present Value (or NPV) of the project.

Year	Cash Flow	PV 10%	Discounted Cash Flow
0	–£1,000,000	1.0000	–£1,000,000
1	–£700,000	0.9091	–£636,364
2	–£500,000	0.8264	–413,223
3	–£400,000	0.7513	–£300,526
4	£3,500,000	0.6830	£2,390,547
NPV			+£40,434

The NPV is positive, and so the project is worthwhile.

The rate of interest used in a DCF is called the target rate (also referred to as the criterion rate, the trial rate or simply the discount rate). The target rate can be chosen in one of a number of ways.

1. If the interest rate at which money is borrowed is used the resulting net present value is the profit.
2. A comparative rate can be used (for example the return shown on government stock — gilts). If the net present value is positive then the project shows a greater return than the alternative it is being compared with, and, ignoring risk, is a better investment.
3. If the project is to be paid for with internal funds the DCF can be prepared either:
   (a) using the rate which could be earned in normal business activity (the personal time preference rate); or
   (b) using the rate the money could earn if invested elsewhere (the external opportunity cost).

In either case a positive NPV indicates a worthwhile project.

Note: If preparing a DCF using a spreadsheet such as Microsoft Excel resist the temptation to use the NPV function, it normally works on the assumption that the cash flows are at fixed intervals and that there is no cash flow in period zero.

The target rate will obviously have a significant effect on the resulting NPV. The following table shows how the NPV of the cash flows above varies with the target rate:

Target Rate	NPV	Target Rate	NPV	Target Rate	NPV
1%	£791,978	11%	–£23,360	21%	–£513,033
2%	£689,671	12%	–£83,996	22%	–£550,088
3%	£592,738	13%	–£141,647	23%	–£585,409
4%	£500,861	14%	–£196,476	24%	–£619,086
5%	£413,742	15%	–£248,638	25%	–£651,200
6%	£331,105	16%	–£298,274	26%	–£681,831
7%	£252,689	17%	–£345,520	27%	–£711,052
8%	£178,254	18%	–£390,504	28%	–£738,934
9%	£107,573	19%	–£433,343	29%	–£765,542
10%	£40,434	20%	–£474,151	30%	–£790,939

Normally, the higher the target rate the lower the NPV (just as with the investment method the higher the all risks yield the lower the market value).

The target rates and resulting NPVs are shown plotted on a graph in Figure 16.1.

Figure 16.1 Graph showing internal rate of return

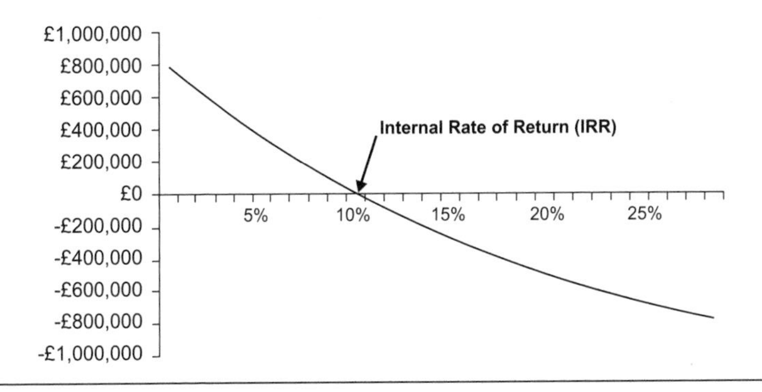

The point where the curve crosses the horizontal axis is the internal rate of return (IRR), that is, the rate which if used to discount the cash flow would give an NPV of zero, in this case about 10.6%.

It is not a practical proposition to calculate a large number of NPVs manually and draw a graph to find an IRR, neither is it practicable to find the IRR by a trial and error.

Spreadsheets like Microsoft Excel offer a simple solution. While the IRR function has the same problems as the NPV function, goal seek (in MS Office 2007 look on the Data tab for What-If Analysis, Goal Seek, in older versions click on Tools, Goal Seek) can be used to make the spreadsheet find the IRR by trial and error. Set the cell with the NPV to 0 by changing the cell which holds the trial rate.

Year	Cash Flow	PV 10.63%	Discounted Cash Flow
0	–£1,000,000	1.0000	–£1,000,000
1	–£700,000	0.9039	–£632,752
2	–£500,000	0.8171	–£408,546
3	–£400,000	0.7386	–£295,438
4	£3,500,000	0.6676	£2,336,735
NPV =			£0
IRR =			10.63%

Alternatively the IRR can be found manually by linear interpolation. Linear interpolation requires the NPVs of the cash flow to be calculated using two different target rates which bracket the IRR (so that one NPV will be positive and the other negative).

The first trial rate can be guessed and its NPV calculated. A second trial rate which will bracket the target can then be estimated from the NPV resulting from the first calculation. If the first NPV was positive the trial rate must be increased, while if the first NPV was negative the trial rate must be decreased (the simple way to remember this is to follow the sign of the first NPV).

Linear interpolation assumes that the curve shown by plotting the NPV values between the two chosen points (shown in Figure 16.2 as a dashed line) is a straight line.

Triangle ACD is similar to triangle ABE, therefore AB/AC = BE/CD.

Therefore (AB/AC) × CD = BE.

BE plus the lowest trial rate (point B on Figure 16.2) is an approximation of the IRR (point E).

Figure 16.2 Using trial rates

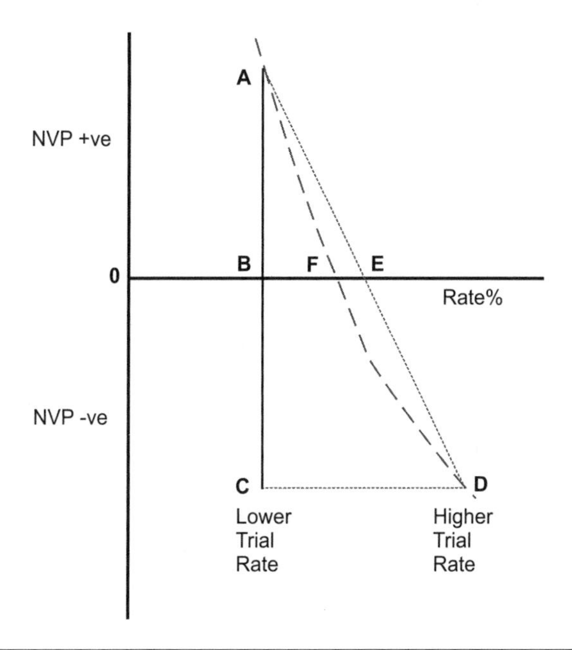

The IRR = the Lower Trial Rate + the difference in the Trial Rates × (the NPV at the Lower Rate/the change in NPV).

The result is an approximation because the curve is not a straight line (the actual IRR is shown as point F on Figure 16.2). The closer the trial rates are to the IRR the more exact the answer will be. If a more accurate figure is required the process can be repeated with successively closer trial rates based on the first approximation of the IRR, although for most practical purposes the first approximation is normally sufficient, particularly bearing in mind the very sensible reluctance of many valuers to use yields which are not expressed in whole, half or sometimes quarter percentage points.

With the sample cash flow the IRR must lie between 10% and 11%, so that these rates can be used to give a good estimate of the IRR:

Year	Cash Flow	Trial Rate 10% PV	DCF	Trial Rate 11% PV	DCF
0	–£1,000,000	1.0000	–£1,000,000	1.0000	–£1,000,000
1	–£700,000	0.9091	–£636,364	0.9009	–£630,631
2	–£500,000	0.8264	–£413,223	0.8116	–£405,811
3	–£400,000	0.7513	–£300,526	0.7311	–£292,477
4	£3,500,000	0.6830	£2,390,547	0.6587	£2,305,558
NPV			£40,434		–£23,360

The IRR = the Lower Trial Rate + the difference in the Trial Rates × (the NPV at the Lower Rate/the Change in NPV)

$$IRR = 10 + 1 \times (40{,}434/(40{,}434 + 23{,}360)) = 10.634\%$$

Both the NPV and the IRR can be used by valuers. The NPV is a 'value' in just the same way as the result of any valuation calculation is a value, while the IRR is a yield, which means that discounted cash flow technique can be a potent system for analysis.

Negative IRRs

It is possible to have a negative Internal Rate of Return. If in the example above the positive cash flow at the end of year four was only £2,500,000 the IRR is –1.354%.

Year	Cash Flow	PV –1.354%	DCF
0	–£1,000,000	1.0000	–£1,000,000
1	–£700,000	1.0137	–£709,610
2	–£500,000	1.0276	–£513,822
3	–£400,000	1.0418	–£416,701
4	£2,500,000	1.0560	£2,640,133
NPV			£0
IRR			–1.354%

Such investment opportunities are best avoided!

Note: The PV figures are greater than 1 where the rate of interest is negative.

The double IRR problem

Occasionally a cash flow may be found which gives two internal rates of return, or which has no IRR. As an example of this phenomenon, the following cash flow gives IRRs of 10% and 40%:

		Trial Rate 10%		Trial Rate 40%	
Year	**Cash Flow**	**PV 10%**	**DCF**	**PV 40%**	**DCF**
0	–£1,000,000	1.0000	–£1,000,000	1.0000	–£1,000,000
1	£2,500,000	0.9091	£2,272,727	0.7143	£1,785,714
2	–£1,540,000	0.8264	–£1,272,727	0.5102	–£785,714
NPV			£0		£0

Figure 16.3 The double IRR problem

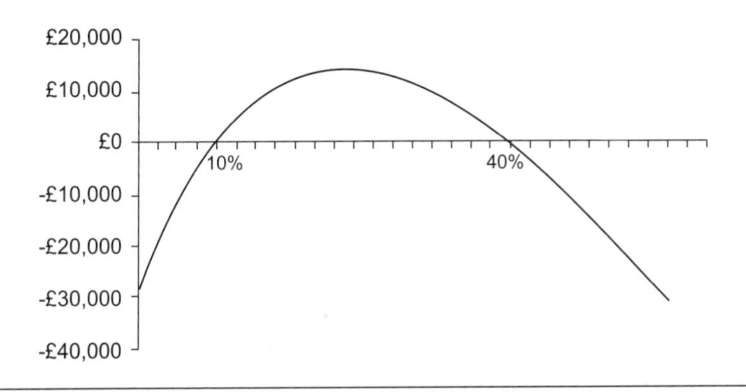

Bouncers can occur when the signs of the cash flows change frequently, which is fortunately unlikely to happen with the cash flows produced by property investments.

Choosing between projects

Discounted cash flow techniques can be particularly useful when faced with a choice between projects by providing data in financial terms on which an informed decision can be based.

Imagine that an investor has a choice of two mutually exclusive schemes with the following cash flows:

Year	Project A	Project B
0	–£1,000,000	–£1,000,000
1	–£430,000	–£2,000,000
2	–£740,000	–£800,000
3	+£3,000,000	+£5,000,000

The first comparison that can be made is to use the NPV method. Let us assume that the investor has a target rate of 10% for projects of this sort. The discounted cash flows are:

Project A

Year	Cash Flow	PV 10%	DCF
0	–£1,000,000	1.0000	–£1,000,000
1	–£430,000	0.9091	–£390,909
2	–£740,000	0.8264	–£611,570
3	£3,000,000	0.7513	£2,253,944
NPV			£251,465

Project B

Year	Cash Flow	PV 10%	DCF
0	–£1,000,000	1.0000	–£1,000,000
1	–£2,000,000	0.9091	–£1,818,182
2	–£800,000	0.8264	–£661,157
3	£5,000,000	0.7513	£3,756,574
NPV			£277,235

So project B has a higher NPV, and is the most profitable.

Alternatively we can compare the internal rates of return shown by the projects. Because the IRRs are unknown, but are above 10% (each project shows a positive NPV with a 10% discount rate) the first trial rate will be set at 15%.

Project A

		Trial Rate 15%		Trial Rate 20%*	
Year	**Cash Flow**	**PV 15%**	**DCF**	**PV 20%**	**DCF**
0	–£1,000,000	1.0000	–£1,000,000	1.0000	–£1,000,000
1	–£430,000	0.8696	–£373,913	0.8333	–£358,333
2	–£740,000	0.7561	–£559,546	0.6944	–£513,889
3	£3,000,000	0.6575	£1,972,549	0.5787	£1,736,111
NPV			+£39,089		–£136,111

* because the NPV at the first trial rate (15%) is positive the second trial rate (20%) should be greater than the first to ensure a negative NPV is used to bracket the target.

$$\text{IRR} = \text{Lower rate} + \Delta \text{ rate} \times \text{NPV @ lower rate}/\Delta \text{ NPV}$$

(Δ is the Greek letter Delta, meaning 'change in').

$$\text{IRR} = 15\% + 5\% \times (£39{,}089/(£39{,}089 + £136{,}111)) = 16.1156\%$$

Using MS Excel's goal seek function the IRR is 16.036%. The difference is caused by approximating the curve as a straight line.

The IRR of Project A is say 16%.

Project B

		Trial Rate 15%		Trial Rate 10%*	
Year	**Cash Flow**	**PV 15%**	**DCF**	**PV 10%**	**DCF**
0	–£1,000,000	1.0000	–£1,000,000	1.0000	–£1,000,000
1	–£2,000,000	0.8696	–£1,739,130	0.9091	–£1,818,182
2	–£800,000	0.7561	–£604,915	0.8264	–£661,157
3	£5,000,000	0.6575	£3,287,581	0.7513	£3,756,574
NPV			–£56,464		+£277,235

* the NPV at the first trial rate is negative, the trial rate must therefore be reduced.

$$\text{IRR} = \text{Lower rate} + \Delta \text{ rate} \times \text{NPV @ lower rate}/\Delta \text{ NPV}$$

$$\text{IRR} = 10\% + 5\% \times (£277{,}235/(£277{,}235 + £56{,}464)) = 14.154\%$$

Using MS Excel's goal seek function the IRR is 14.078%, again the difference is caused by the approximation of the curve as a straight line.

The IRR of Project B is say 14%.

Project A, with an IRR of 16%, is better than Project B (IRR = 14%).

There is therefore a problem in advising the imaginary client. Each project plainly has advantages. Project B produces a greater capital profit, while Project A gives the higher percentage return on the money invested.

In most cases the investor should opt for the choice which gives the greatest monetary return, and the NPV approach is now commonly the only one considered.

There are other checks which can help to prevent schizophrenia.

Incremental analysis

Incremental analysis involves deducting the cash flows of the project with the lower outlay from the cash flows of the project with the higher outlay. The IRR of the incremental cash flows can then be found:

Year	Project B Cash Flow	Project A Cash Flow	Incremental Cash Flow	PV 10.97%	DCF
0	–£1,000,000	–£1,000,000	£0	1.0000	£0
1	–£2,000,000	–£430,000	–£1,570,000	0.9011	–£1,414,773
2	–£800,000	–£740,000	–£60,000	0.8120	–£48,722
3	£5,000,000	£3,000,000	£2,000,000	0.7317	£1,463,495
NPV					£0

The incremental yield (10.97%) is greater than the target rate (10%), showing that the increased expenditure is at least yielding more than the cost of finance. In these circumstances the investor should, assuming the risks are equal, choose project B, as while it shows a lower percentage return than project A, it uses more capital profitably in achieving its higher capital return.

Benefit cost ratio

The benefit cost ratio is the discounted present value of the total benefits of a project divided by the discounted present value of the total costs. This method is only of use when the target rate of the projects is not known, when the project with the highest benefit cost ratio should be chosen, as the high cost low yield project might show a return less than the investor's target yield.

Years Purchase is DCF

When a cash flow table has the same amounts of income or expenditure in a number of successive years a YP can be used to get a present value of the combined flows, thus saving some rather tedious arithmetic using PVs.

The YP (or Present Value of £1pa) can be found by one of the following formulae:

$$YP = (1 - PV)/i$$

or

$$YP = 1/(i + SF)$$

Where:

i = the interest rate expressed as a decimal
PV = the Present Value of £1 = $1/(1 + i)^n$
SF = the Annual Sinking Fund = $i/(1/(1 + i)^n - 1)$
n = the number of years.

If the cash flow is the same amount per annum for ever the YP in perp is used.

$$YP \text{ in perp} = 1/i$$

So, assuming a target rate of 10%, the net present value of a cash flow of £75,000pa for three years followed by a reversion to £100,000pa in perpetuity can be found in one of the following ways (in these examples all the decimal places provided by a pocket calculator have been used, although only four decimal places are shown in the calculations).

1. As a short cash-flow using YPs

Figure 16.4 Short cash flow

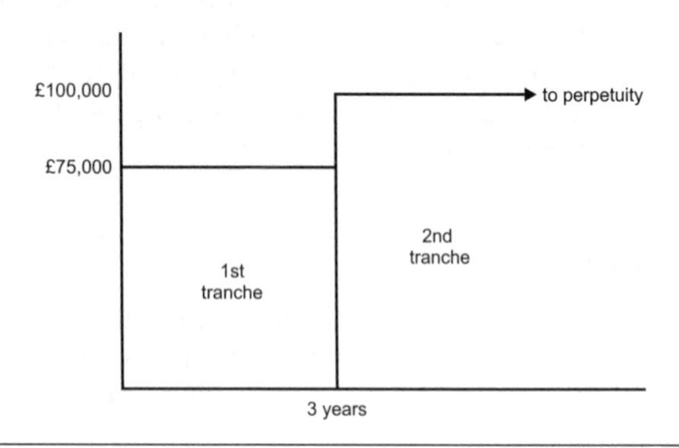

Years	Cash Flow	YP 10%	PV 10%*	DCF
1–3	£75,000	2.4869	1.0000	£186,514
4 +	£100,000	10.0000	0.7513	£751,315
NPV				£937,829

* Notice that the PVs are for the period before the cash flow (that is in the first line for 0 years not 1, and in the second line, for 3 years not 4, so 0.7513 is the PV of £1 in 3 years). This is because the YP brings the cash flow back to the start of the year.

2. As a longer cash flow using only PVs

Figure 16.5 Longer cash flow

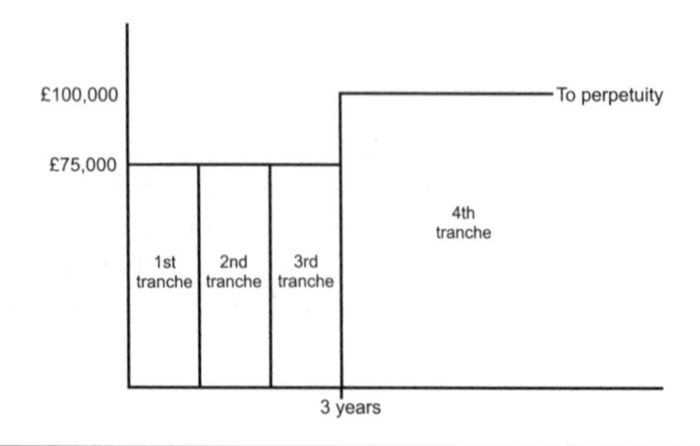

Year	Cash Flow	PV 10%	DCF
1	£75,000	0.9091	£68,182
2	£75,000	0.8264	£61,983
3	£75,000	0.7513	£56,349
4 +	£100,000	7.5131*	£751,315
NPV			£937,829

* YP in perp deferred 3 years, turning the annual cash flow into a capital sum and deferring it.

3. As a traditional 'term and reversion' investment method valuation

Figure 16.6 Traditional term and reversion

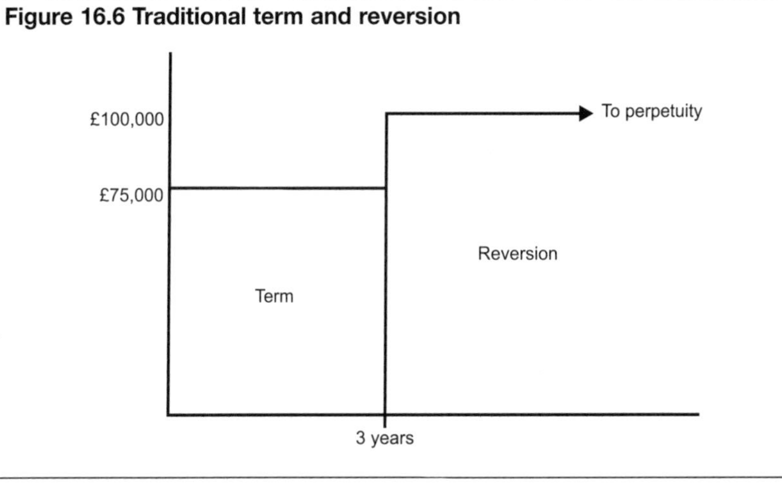

Term (3 years)		
Net income	£75,000pa	
× YP 3 years at 10%	2.4869	
		£186,514
Reversion		
Net income	£100,000pa	
× YP in perp deferred 3 years at 10%	7.5131	
		£751,315
		£937,829

4. As a 'hardcore' investment method valuation

Figure 16.7 Hardcore

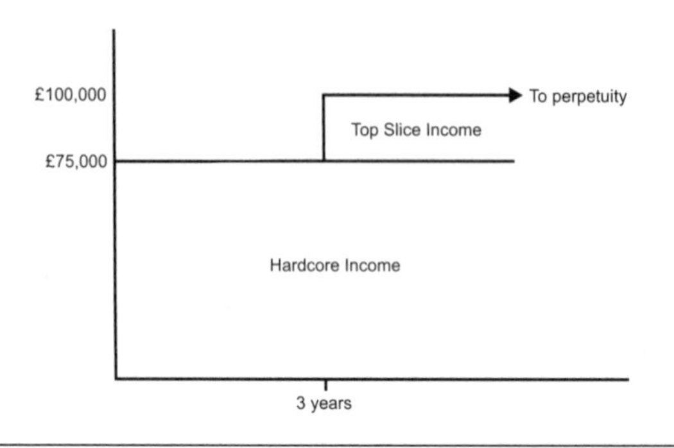

Hardcore income	£75,000pa	
× YP in perp 10%	10	
		£750,000
Top slice	£25,000pa	
× YP in perp deferred 3 years at 10%	7.5131	
		£187,829
		£937,829

If nothing else, this series of calculations demonstrates that discounted cash flow and the investment method of valuation covered in Chapter 14 use the same mechanics, but with a different layout. In other words, the traditional investment method approaches are in fact discounted cash flow valuations.

The DCF format can however be more tedious to use than the more usual well tried investment method valuation. Apart from the fact that discounted cash flow is very widely accepted for all kinds of investment appraisal, what does it offer to the valuer?

Discounted cash flow is more flexible than the traditional valuation layouts. It copes easily with unusual income flows, and forces the valuer to specify explicitly the income flows and patterns being dealt with, and can take full account of the timing of payments. This means that property, let on a historic lease with terms requiring the payment of rent on a basis which is no longer usual in the market place, can be valued without either ignoring the problems inherent in the valuation, or recourse to complex arguments on how to deal with income annually in arrears when the yield structure is based on analysis of property let quarterly in advance.

Analysis

One advantage of DCF is in analysis. We have already seen that the internal rate of return can be used to compare investments. The equivalent yield (see Chapter 14), widely used by valuers, is an IRR, and in many cases the most efficient way to find it is by using DCF techniques.

Exactly the same method can be used to examine the yields of dated government stocks (gilts), which are government IOUs. For example, an 8% stock with a redemption date in five years sells for £118.81 per £100 nominal value.

The interest only yield is £8/£118.81 = 6.733%

However this yield does not reflect the £100 face value which will be repaid to the stock holder in five years. The redemption yield which takes the redemption of the stock into account, together with the fact that interest on most gilts is paid twice a year, can be calculated as follows:

		Trial Rate 4%		**Trial Rate 3.75%**	
Years	**Cash Flow**	**PV 4%**	**DCF**	**PV 3.75%**	**DCF**
0	–£118.81	1.0000	–£118.81	1.0000	–£118.81
0.5	£4.00	0.9806	£3.92	0.9818	£3.93
1	£4.00	0.9615	£3.85	0.9639	£3.86
1.5	£4.00	0.9429	£3.77	0.9463	£3.79
2	£4.00	0.9246	£3.70	0.9290	£3.72
2.5	£4.00	0.9066	£3.63	0.9121	£3.65
3	£4.00	0.8890	£3.56	0.8954	£3.58
3.5	£4.00	0.8717	£3.49	0.8791	£3.52
4	£4.00	0.8548	£3.42	0.8630	£3.45
4.5	£4.00	0.8382	£3.35	0.8473	£3.39
5	£104.00	0.8219	£85.48	0.8319	£86.52
NPV			–£0.65		£0.58

IRR = Lower rate + Δ rate × NPV @ lower rate/Δ NPV

IRR = 3.75% + 0.25% × 58p/(58p + 65p) = 3.87% pa

It is possible to derive years purchase figures for half years and so reduce the length of the calculation. A simpler approach is to regard each half year as a year and use the normal valuation tables:

		Trial Rate 2%			Trial Rate 1.75%		
Years	Cash	PV	YP	DCF	PV	YP	DCF
0	–£118.81	1.0000		–£118.81	1.0000		–£118.81
1–9	£4.00	1.0000	8.1622	£32.65	1.0000	8.2605	£33.04
10	£104.00	0.8203		£85.32	0.8407		£87.44
				–£0.84			£1.67

$$\text{IRR} = \text{Lower rate} + \Delta\ \text{rate} \times \text{NPV @ lower rate}/\Delta\ \text{NPV}$$

$$\text{IRR} = 1.75\% + 0.25\% \times £1.67/(£1.67 + £0.84) = 1.9163\% \text{ per half year}$$

$$\text{IRR on an annual basis} = ((1+ 0.019163)^2 - 1) = 3.87\% \text{ pa}$$

Discounted cash flow tables can also be prepared which include an allowance for growth, so that the calculation explicitly reflects the impact of inflation. An IRR found in this way is an equated yield, which allows equity investments (that is an investment with growth potential, including most property interests) to be compared readily with fixed interest securities (such as government stock). Chapter 17 outlines this approach.

Further reading

The Income Approach to Property Valuation, Baum A, Mackmin D and Nunnington N, 5th ed, EG Books, 2006, particularly Chapter 4.

17 Contemporary Growth Explicit Methods of Valuation

"For many decades the conventional methods of investment valuation were accepted as logical, practical, and seemingly immutable". *Property Valuation Methods Interim Report*, prepared for the RICS by the Polytechnic of the South Bank, 1980

This is not to say that conventional methods have not been challenged. This chapter explores alternative, growth explicit, valuation methods and examines the way in which they can address some of the problems encountered within traditional growth implicit methods.

Freehold valuation and growth

Chapter 14 looked at the traditional approach to the valuation of freehold interests using the investment method, and demonstrated the way in which the yield depended on the relative security or risk attached to an income flow.

It was explained that one of the key risks was whether the income flow was *inflation proof* (because it had growth potential as an equity investment) or *inflation prone* (a fixed interest security). An inflation prone income will decline in real terms over time while an inflation proof income flow should keep pace with inflation.

Relatively high discount rates are used to value inflation prone income flows because they have a high risk of falling in value in real terms, and relatively low discount rates are used for inflation proof income flows because they offer a hedge against inflation, and hence a lower risk in real terms.

The traditional approach uses the all risks yield to capitalise the income in the reversion in order to take account of future growth potential, and it is argued that there is no need to consider future growth separately. The rule is as you devalue, so shall you value and the all risks yield should have been found by analysing comparable transactions without making an explicit assumption about future income growth. If the comparables are good the growth potential will be the same as that for the subject property — the valuer is not ignoring growth potential, it is implicitly taken into account in the yield.

However valuers make intuitive adjustments to the all risks yield to reflect changes in risk. So, in Example 14.6 the yield for the term was reduced by 0.25% to reflect extra security of income, and it was noted that the expected growth in market rent was accounted for in the yield adopted to value the reversion, the lower the yield the higher the level of anticipated growth. In Example 14.8 the yield used to value the term was increased by 2% because the income was fixed for the 20 years of the term and so was inflation prone.

While these adjustments are logical, they are difficult to support empirically, and the practice of reducing the yield used for the valuation of the term has attracted considerable criticism. First, while the income might be more secure because there is a limited risk of tenant default, it is not secure in real terms because there is no growth potential. Second, reducing the yield by a fraction of a percent can increase the value of the interest to more than that which would be shown if the property had been recently let at the market rent.

It is possible to prepare valuations that take growth potential into account by using the right yield for each income flow.

Figure 17.1 shows three ways in which an income flow can change with time.

The dotted line shows a cash flow where the income has the potential to grow all the time. This occurs if a property is vacant — the market rent which it could be let for has the capacity to grow immediately. Such an income should be valued using the Inflation Risk Free Yield — the yield which the property investment would show if growth occurred immediately, rather than at periodic rent reviews.

The inflation risk free yield can be found using the following formula.

Figure 17.1 How incomes can change over time

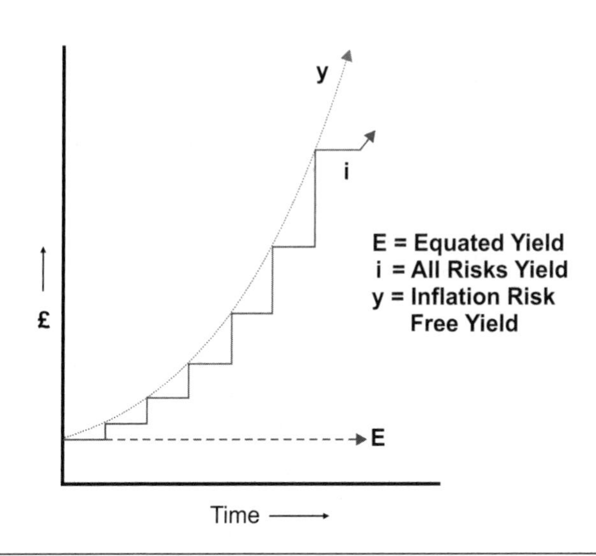

$$\text{Inflation Risk Free Yield} = \left[\frac{\text{Equated Yield} + 1}{\text{Growth Rate} + 1}\right] - 1$$

The dashed line in Figure 17.1 shows an income which is fixed, with no ability to grow. It is inflation prone. A cash flow like this should be valued using the *Equated Yield* — the internal rate of return which when applied to the projected income flow results in the sum of the incomes discounted at that rate equalling the capital value. Rents at review, lease renewal or re-letting take into account expected future rental growth.

For valuation purposes the equated yield can be fixed by investor's requirements, or it can be derived from alternative investments such as gilts (government securities). This can be used to assess a risk free return, to which a property investment risk premium is added.

An addition of 2% for a property investment risk premium seems to be common, however in "Estimation and Analysis of the Risk Premium for Commercial Property" (*Journal of Property Investment and Finance*, vol 17, no 3, 1999) Heather Tarbert and John-Paul Marney conclude that "the 2% traditional required risk premium over conventional gilts is not stable and is not realised if earlier time periods (pre 1978) are discarded ... The historic risk premium over conventional gilts appears to fluctuate around a value of zero ... The risk premium is negative if the comparison is against index linked gilts".

Given the returns on gilts at the time of writing an equated yield of around 8% might be appropriate (see Chapter 18 for a detailed explanation).

The solid stepped line in Figure 17.1 shows the cash flow from an inflation proof property investment where the income will increase at each rent review. This income should be valued using an all risks yield which reflects implied growth. The all risks yield varies with the review pattern — a short rent review pattern gives a lower risk of the income being eroded by inflation, and so will show a lower all risks yield.

If the equated yield and growth rate are known the all risks yield can be found using the following formula:

$$\text{all risks yield} = i = E - E\left[\frac{(1+g)^n - 1}{(1+E)^n - 1}\right]$$

Where:

i = all risks rate
E = equated yield
g = growth rate
n = number of years

Growth rate

Having established what yield to use for each cash flow we need to be able to find the growth rate.

The growth rate required is not the historic growth rate in rents (which can be easily found by analysing past records) because growth in the past may well be different from the growth that will occur in the future. If the equated yield and the all risks rate are known the anticipated future growth in market rent can found using the following formula:

$$g = \left[\sqrt[n]{1 + \frac{E - i}{\text{SF } E\% \text{ } n \text{ years}}}\right] - 1$$

Where:
i = all risks rate
E = equated yield
g = growth rate
n = number of years
SF = annual sinking fund

If there are annual rent reviews:

Equated Yield – All Risks Yield = Implied Growth Rate

Equated yield, all risks yield and the implied growth rate form a triangle. If two are known the other can be found. The all risks rate is normally known, the equated yield is then assumed and the implied growth rate derived from them.

The table below shows the implied growth rates possible for a range of all risk and equated yields given a five year review pattern:

5 yr reviews	Equated Yield						
All risks yield	7%	8%	9%	10%	11%	12%	13%
5.0%	2.20%	3.30%	4.39%	5.47%	6.56%	7.64%	8.71%
6.0%	1.12%	2.24%	3.36%	4.47%	5.57%	6.67%	7.77%
7.0%	0.00%	1.15%	2.29%	3.42%	4.55%	5.67%	6.79%
8.0%	–1.18%	0.00%	1.17%	2.33%	3.49%	4.63%	5.77%
9.0%	–2.41%	–1.20%	0.00%	1.19%	2.38%	3.55%	4.72%
10.0%	–3.72%	–2.47%	–1.23%	0.00%	1.22%	2.42%	3.62%

If the rent review pattern is three years the figures become:

3 yr reviews	Equated Yield						
All risks yield	7%	8%	9%	10%	11%	12%	13%
5.0%	2.10%	3.15%	4.19%	5.24%	6.28%	7.32%	8.37%
6.0%	1.06%	2.12%	3.18%	4.23%	5.29%	6.34%	7.39%
7.0%	0.00%	1.07%	2.14%	3.21%	4.27%	5.33%	6.40%
8.0%	–1.08%	0.00%	1.08%	2.16%	3.24%	4.31%	5.38%
9.0%	–2.19%	–1.09%	0.00%	1.09%	2.18%	3.27%	4.35%
10.0%	–3.32%	–2.21%	–1.10%	0.00%	1.10%	2.20%	3.30%

Valuations on a growth explicit basis

Example 17.1

The reversionary freehold in Example 14.6 from Chapter 14 will be used to demonstrate the *Real Value Equated Yield Hybrid Valuation*, which was developed in the 1980s by Crosby from Wood's Real Value Approach of the 1970s.

Value the freehold interest in a modern office block. One year ago the office was let to a government department at £80,000pa. The lease contains a clause to review the rent every five years. The market rent is £100,000pa.

The all risks rate adopted in the example was 5%, presumably found by analysis of comparable property investment sales. If the appropriate equated yield is say 8%, the implied growth rate is:

$$g = \left[\sqrt[n]{1 + \frac{E - i}{\text{SF } E\% \text{ } n \text{ years}}} \right] - 1$$

Where:

i = all risks rate = 5% or 0.05
E = equated yield = 8% or 0.08
g = growth rate
n = number of years = 5 (that is the rent review period)
SF = annual sinking fund 8% 5 years = 0.1704565

$$g = \left[\sqrt[5]{1 + \frac{0.08 - 0.05}{0.1704565}} \right] - 1 = 0.0329548 \text{ or } 3.29548\%$$

The valuation can now be prepared (in this example all the decimal places provided by a pocket calculator have been used, although only four decimal places are shown in the calculations):

Valuation of freehold interest in office block using real value equated yield hybrid

Term		
Rent reserved FRI	£80,000pa	
× YP 4 years at 8%	3.3121	
(The equated (no growth) yield is used as there is no growth potential during the 4 year term in the £80,000pa rent)		£264,970

Reversion		
Market rent FRI now	£100,000pa	
Allow for growth during the term		
× Amount of £1 for 4 years at the growth rate		
= $(1 + 0.0329548)^4$ =	1.13848	
Future projected market rent on review	£113,848	
× YP in perp at 5%		
(The all risks rate is used because the forecast reversionary income of £113,848pa has future growth potential)	20	
Capital value of reversion in four years	£2,146,869	
(The capital value of the reversion must then be deferred to give the value now, rather than the value when the rent review occurs. There is no growth potential unaccounted for in the 'waiting period', that is the term, and so the equated yield is used)		
× PV in 4 years at 8%	0.7350	
		£1,673,633
Value of freehold interest		£1,938,603

Example 17.2

An alternative mathematically identical method, which might be preferred by valuers who are unhappy about making an explicit projection of a future market rent, uses the current market rent but defers the reversion at the inflation risk free yield because the market rent has the capacity to grow all the time between the valuation date and the rent review date.

$$\text{Inflation Risk Free Yield} = \left[\frac{\text{Equated Yield} + 1}{\text{Growth Rate} + 1}\right] - 1$$

$$\text{Inflation Risk Free Yield} = \left[\frac{1 + 0.08}{1 + 0.0329548}\right] - 1 = 0.0455443 \text{ or } 4.5544\%$$

Valuation of freehold interest in office block using alternative real value approach:

Term		
Rent reserved FRI	£80,000pa	
× YP 4 years at 8%	3.3121	
(The equated yield is used as there is no growth potential)		£264,970
Reversion		
Market rent FRI now	£100,000pa	
× YP in perp at 5% (The all risks rate is used because the income has growth potential)	20	
Capital value of reversion in four years	£2,000,000	
(The capital value of the reversion must then be deferred to give the value now using the inflation risk free yield because the current market rent has the capacity to grow)		
× PV in 4 years at 4.5544%	0.8368	
		£1,673,633
Value of freehold interest		£1,938,603

Comparing the results from the growth explicit valuations with the valuation figure produced by the traditional approach the final valuation figures are almost the same, and the difference would be lost in rounding, however the values for the term and the reversion which make up the valuations show significant differences:

	Traditional	**Growth Explicit**	**Difference**
Term	£285,331	£264,970	+7.68%
Reversion	£1,645,405	£1,673,633	–1.69%
Total	£1,930,736	£1,938,603	–0.40%

The traditional approach overvalues the term because the yield used assumes the fixed income flow has growth potential when it does not. It then undervalues the reversion by using the current market rent without allowing for growth between the valuation date and the end of the term. Provided the reversion is close and the rent passing is in the same order as the market rent (which will be true if the property is let on a modern lease structure) these errors tend to cancel each other out.

The fact that, in most cases, the traditional approach's combination of over valuing the term and under valuing the reversion produces about the same answer as a growth explicit valuation may explain why the growth explicit models appear not to have gained any acceptance in the market in the 25 years since they were developed, although they solve many of the problems with traditional valuation methods.

If property is not let with frequent reviews the difference between the two approaches can become more significant, particularly if the intuitive adjustment of the yield used to value the term is not correct.

Example 17.3

Example 14.8 in Chapter 14 considered a freehold interest let on a long lease with 20 years to run without review. The yield for the term was taken at 7%, 2% above the all risks yield, to reflect the lack of growth potential.

The all risks rate adopted in the example was 5%, presumably found by analysis of comparables. If the appropriate equated yield is say 8%, the implied growth rate is:

$$g = \left[\sqrt[n]{1 + \frac{E - i}{\text{SF } E\% \text{ } n \text{ years}}} \right] - 1$$

i	=	all risks rate = 5% or 0.05
E	=	equated yield = 8% or 0.08
g	=	growth rate
n	=	number of years = 5 (the normal review pattern)
SF	=	annual sinking fund 8% 5 years = 0.1704565

$$g = \left[\sqrt[5]{1 + \frac{0.08 - 0.05}{0.1704565}} \right] - 1 = 0.0329548 \text{ or } 3.29548\%$$

Valuation of freehold interest in office block let on long lease using real value equated yield hybrid

Term		
Rent reserved FRI	£80,000pa	
× YP 20 years at 8%	9.8181	
(The equated yield is used as there is no growth potential)		£785,452

Reversion		
Market rent FRI now	£100,000pa	
Allow for growth		
× Amount of £1 for 20 years at the growth rate		
= $(1 + 0.032954)^{20}$ =	1.9126	
Market rent on review	£191,261pa	
× YP in perp at 5%		
(The all risks rate is used because the income has future growth potential)	20	
Capital value of reversion in twenty years	£3,825,219	
(The capital value of the reversion must then be deferred to give the value now, rather than the value when the rent review occurs. There is no growth potential unaccounted for in the 'waiting period', and so the equated yield is used).		
× PV in 20 years at 8%	0.2145	
		£820,694
Value of freehold interest		£1,606,146

Comparing the term and reversion valuations:

	Term	**Reversion**	**Total**
Traditional equivalent yield	£996,977	£753,779	£1,750,756
Traditional yield for term adjusted upwards	£847,521	£753,779	£1,601,300
Growth Explicit	£785,452	£820,694	£1,606,146

The traditional approach again over values the term and undervalues the reversion, and because the value of the longer term is a greater proportion of the value of the whole the effect is now significant. The problem can be corrected if the yield used to value the term in the traditional valuation is increased; the problem is it is very hard to judge by how much.

The hardcore valuation will produce the same answer as the traditional approach if the same yield is used for all parts of the valuation, that is an equivalent yield. Adjusting the yield to value the hardcore income is not a good solution because the top slice, which carries all the growth potential, will be significantly undervalued. The valuer who works on an equivalent yield basis will have to increase the interest rate used, but by how much? Again this is hard to judge.

Growth explicit valuation techniques also provide a rational method for valuing over rented property.

Example 17.4

Imagine that a shop was let two years ago for 15 years at £100,000pa on FRI terms subject to a five yearly upward only rent review. The economy is now in recession, values have fallen and the market rent is now only £90,000pa on the same terms. The all risks rate, by comparison is say 6%.

Valuation of freehold interest in over-rented shop adopting a traditional approach:

Term		
Rent reserved FRI	£100,000pa	
× YP 3 years at say 6.5%	2.6484	
		£264,848
Reversion		
Market rent FRI	£90,000pa	
× YP in perp deferred 3 years at 6%	13.9937	
		£1,259,429
Value of freehold interest		£1,524,277

The yield for the term has been increased from 6% because the income is not secure; if the tenant defaults the rent will drop to market rent, assuming that a new tenant can be found.

In these circumstances it is tempting to adopt a hardcore approach.

Valuation of freehold interest in over-rented shop adopting a hardcore approach:

Bottom slice		
Hardcore income — market rent FRI	£90,000pa	
× YP in perp at 6%	16.6667	
		£1,500,000
Top slice		
Rent reserved FRI	£100,000pa	
less market rent FRI	£90,000pa	
Top slice income	£10,000pa	
× YP 3 years at 7%	2.6243	
		£26,243
Value of freehold interest		£1,526,243

The top slice is a terminating income, so it might be argued that a dual rate approach should be used. In any event the income is not secure, so a higher yield has been adopted. A criticism of the valuation is that we don't know when the market rent will increase to match the rent reserved. The all risks yield at 6% certainly suggests that market anticipates future growth ...

Assuming that an equated yield of 8% is appropriate, the implied growth rate is:

$$g = \left[\sqrt[n]{1 + \frac{E - i}{SF\ E\%\ n\ years}} \right] - 1$$

i = all risks rate = 6% or 0.06
E = equated yield = 8% or 0.08
g = growth rate
n = number of years = 5 (the normal review pattern)
SF = annual sinking fund at 8% for 5 years = 0.1704565

$$g = \left[\sqrt[5]{1 + \frac{0.08 - 0.06}{0.1704565}} \right] - 1 = 0.022437 \text{ or } 2.2437\%$$

In three years time, on the first rent review, the market rent will be
£90,000 × Amount of £1 in 3 years at 2.2437%
= £90,000 × 1.022437^3
= £90,000 × 1.0688 = £96,194pa — less than the £100,000 that the tenant is obliged to pay because of the upward only rent review.

On the second review the market rent will be:
£90,000 × Amount of £1 in 8 years at 2.2437%
= £90,000 × 1.022437^8
= £90,000 × 1.1942 = £107,482pa — so the rent can be increased.

The valuation becomes:

Valuation of freehold interest in over-rented shop using Real Value Equated Yield Hybrid

Term		
Rent reserved FRI	£100,000pa	
× YP 8 years at 8%	5.7466	
(The equated yield is used as there is no growth potential)		£574,664

Reversion		
Market rent FRI now	£90,000pa	
Allow for growth		
× Amount of £1 for 8 years at the growth rate		
× 1.022437^8	1.1942	
Market rent on review	£107,482pa	
× YP in perp at 6%		
(The all risks rate is used because the income has future growth potential)	16.6667	
Capital value of reversion in four years	£1,791,360	
(The capital value of the reversion must then be deferred to give the value now, rather than the value when the rent review occurs. There is no growth potential unaccounted for in the 'waiting period', and so the equated yield is used).		
× PV in 8 years at 8%	0.5403	
		£967,816
Value of freehold interest		£1,542,480

If the market is using the traditional or hardcore approaches demonstrated above the value is slightly less than this, the owner might wish to retain the interest as it is worth more than the price it is likely to achieve. In some circumstances the position may be reversed, with the value higher than the worth to the owner, in which case the owner should consider selling.

"There continues to be a widespread belief among practitioners that existing valuation methods are sacrosanct and that because they have always been used they should continue to be used. It is argued that changes are not made in practice despite this theoretical justification, because 'the market does not value that way'. This presupposes that valuers are accurately interpreting and reflecting market behaviour, a questionable assumption ... unless the surveying profession is prepared to respond positively to such innovation, the consequences are likely to be far-reaching and unfavourable" (Trott "Property Valuation Methods — Research Report", RICS, 1986, quoted in "Current Valuation Practice in Australia, A Survey of Valuers", Terry Boyd, *Journal of Property Valuation & Investment*, vol 13, no 3, 1995 pp 59–69).

As long as valuers continue to use the traditional approach, with occasional intuitive adjustments to deal with unusual lease lengths, a growth explicit calculation can be used to find the worth of the investment, and help investors decide whether to buy or sell. If this becomes common the market price will tend to follow the growth explicit calculation, and the traditional approach will become redundant.

Leasehold valuation and growth

Chapter 15 looked at the traditional approach to leasehold valuation, which involved finding the profit rent and then capitalising it using an appropriate YP.

The major difficulty is that capitalising a profit rent is not a good way to value leasehold interests because the profit rent does not reflect the growth potential, which can be anything from highly geared to nil, see Example 17.5 below.

When growth is taken into account a profit rent might be one of four possible alternative potential income flows. In each case the profit rent is the same, and the value of the leasehold interest is represented by the shaded area in the diagram.

1. The rent collected will grow, while the rent paid is fixed

Figure 17.2 Lease valuation and growth

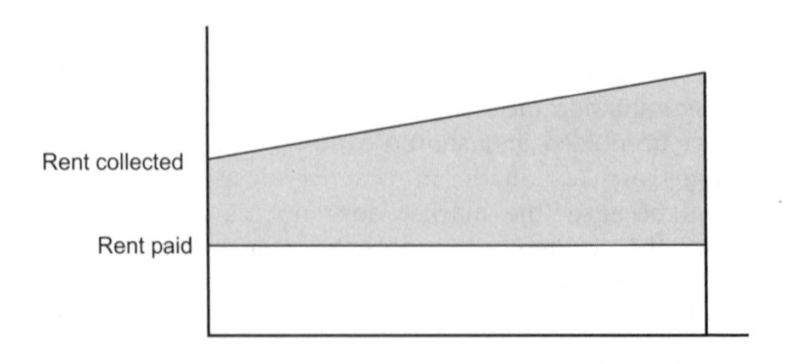

This is the most common situation, a tenant occupied property, where the rent collected in theory is the market rent, able to rise

all the time, or a property sub-let on a short lease or with regular rent reviews, where the rent collected will rise in steps at the end of each lease or on each rent review.

2. The rent collected will grow, as will the rent paid

Figure 17.3 Growth in rent paid

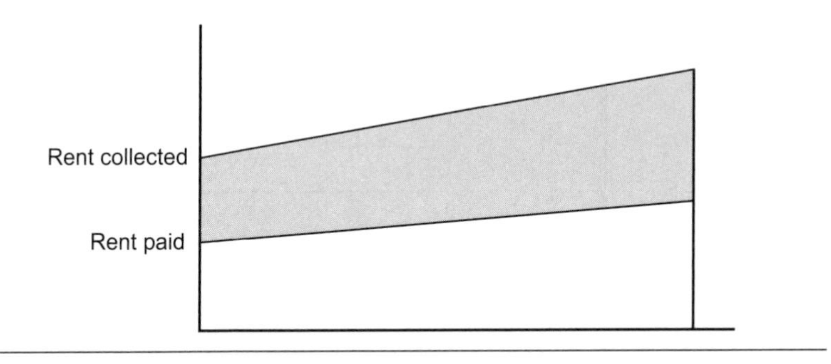

Found on some modern commercial ground leases. The freeholder benefits from regular rent reviews, often to a percentage of either the rent collected or the market rent.

3. The rent collected and the rent paid are both fixed

Figure 17.4 Fixed rents

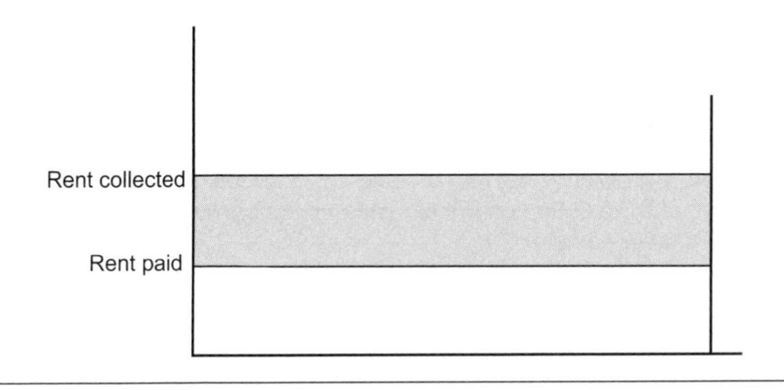

A sub-letting at a fixed rent for the remainder of the head-leaseholder's term. This is found where a developer holds a site on a long lease and sells one or more long leasehold interests in the completed development.

4. The rent collected is fixed, while the rent paid will rise.

Figure 17.5 Negative gearing

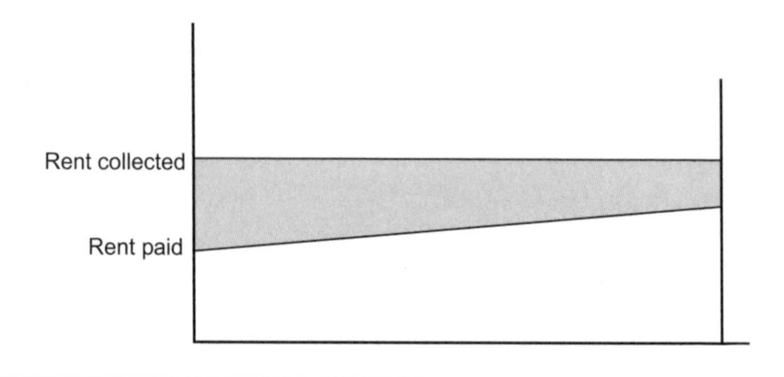

Hopefully not found, a high risk situation for the head-lessee because there is a risk that the rent paid will grow to more than the rent collected.

While valuers may suggest that intuitive adjustments can be made to the yield to reflect the growth potential of the profit rent (or lack of it), the cash flows shown by leasehold interests when either the rent paid or the rent collected has growth potential are complex.

Example 17.5

Consider two leasehold interests in similar properties, each with 20 years to run at fixed rents. Leasehold A has a market rent of £50,000pa on FRI terms assuming five year reviews. The rent payable is £25,000pa net. Leasehold B has a market rent of £125,000pa on FRI terms assuming five year reviews. The rent payable is £100,000pa net.

Traditional investment method valuations of the two interests would start:

	Leasehold A	**Leasehold B**
Market rent FRI (5 year reviews)	£50,000pa	£125,000pa
less rent paid (same terms)	£25,000pa	£100,000pa
Profit Rent	£25,000pa	£25,000pa

And would then capitalise the profit rent by a YP — however this is derived, whether single rate, dual rate, or dual rate tax adjusted, the values of Leasehold A and Leasehold B will be the same. But the two interests have different growth patterns.

Let us assume a growth rate of say 2%. The growth factor for each period can be found using the Amount of £1 for five years at 2%. The cash flows from the interests are:

Leasehold A

Years	**Growth**	**Rent Received**	**Rent Paid**	**Profit Rent**
1 to 5	1	£50,000	£25,000	£25,000
6 to 10	1.1041	£55,205	£25,000	£30,205
11 to 15	1.2190	£60,950	£25,000	£35,950
16 to 20	1.3459	£67,293	£25,000	£42,293

Leasehold B

Years	**Growth**	**Rent Received**	**Rent Paid**	**Profit Rent**
1 to 5	1	£125,000	£100,000	£25,000
6 to 10	1.1041	£138,010	£100,000	£38,010
11 to 15	1.2190	£152,374	£100,000	£52,374
16 to 20	1.3459	£168,234	£100,000	£68,234

Figure 17.6 below shows these cash flows plotted on a pair of diagrams.

The value of the interests is represented by the shaded area on each diagram. It is obvious that Leasehold B is worth significantly more than Leasehold A. The reason is that profit rents are top slice income. This is commonly thought to be a negative factor, as the income is at higher risk than the freehold hardcore income or bottom slice. However the top slice income picks up all of the growth potential, it is said to be highly geared.

The actual growth of each investment for each five year period can be found by dividing the profit rent by the profit rent for the period before and subtracting 1. The annual growth rate is the fifth root of (1 + the growth rate) –1.

For example on the first review the profit rent for Leasehold A has increased from £25,000 to £30,205, an increase of (£30,205/£25000) – 1 = 0.208 = 20.8% over 5 years.

Figure 17.6 Leaseholds A and B compared

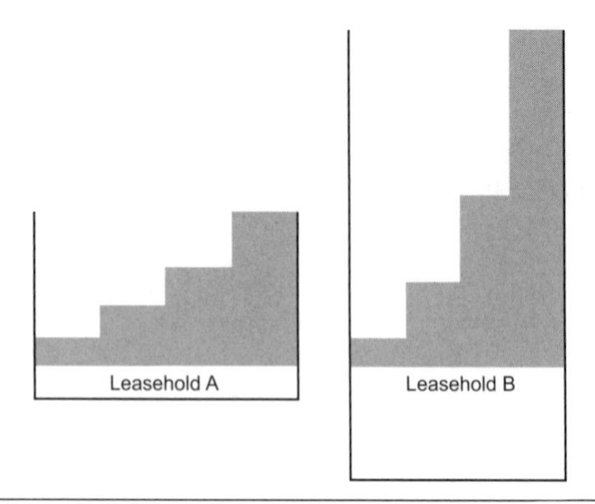

The annual increase is $(\sqrt[5]{1.208}) - 1 = 0.038544 = 3.9\%$.

The results of these calculations are shown below.

Leasehold A

Years	**Profit Rent**	**Growth**	**Growth pa**
1 to 5	£25,000		
6 to 10	£30,205	20.8%	3.9%
11 to 15	£35,950	19.0%	3.5%
16 to 20	£42,293	17.6%	3.3%

Leasehold B

Years	**Profit Rent**	**Growth**	**Growth pa**
1 to 5	£25,000		
6 to 10	£38,010	52.0%	8.7%
11 to 15	£52,374	37.8%	6.6%
16 to 20	£68,233	30.3%	5.4%

A discounted cash flow can then be used to value the two interests. The discount rate is a property based equated yield, which will probably be somewhat higher than that shown by a similar freehold, as leaseholds have traditionally been regarded as higher risk investments (the top slice argument, but once growth is accounted for separately it has more validity). For the purposes of this series of calculations an equated yield of 8% has been assumed.

Leasehold A

Years	Growth	Equated Yield 8% Cash Flow	YP	PV	DCF
1 to 5	1.0000	£50,000	3.9927	1.0000	£199,636
6 to 10	1.1041	£55,204	3.9927	0.6806	£150,010
11 to 15	1.2190	£60,950	3.9927	0.4632	£112,720
16 to 20	1.3459	£67,293	3.9927	0.3152	£84,700
1 to 20	1.0000	–£25,000	9.8181	1.0000	–£245,454
NPV					£301,612

Leasehold B

Years	Growth	Equated Yield 8% Cash Flow	YP	PV	DCF
1 to 5	1.0000	£125,000	3.9927	1.0000	£499,089
6 to 10	1.1041	£138,010	3.9927	0.6806	£375,025
11 to 15	1.2190	£152,374	3.9927	0.4632	£281,801
16 to 20	1.3459	£168,234	3.9927	0.3152	£211,750
1 to 20	1.0000	–£100,000	9.8181	1.0000	–£981,815
NPV					£385,850

The valuation of Leasehold A could be found by capitalising the profit rent at £301,612/£25,000 = 12.06 YP, a yield on a single rate basis of about 5.5%.

For Leasehold B the valuation figure is £385,849/£25,000 = 15.43 YP, a single rate yield of about 2.6%.

So the valuer adopting the traditional approach of capitalising the profit rent has to make significant yield adjustments to arrive at valuation figures which reflect the way in which the growth potential is all in the top slice income.

It can be argued that the example is artificial. Most modern leases have only a few years to run before either the rent paid increases to the market rent on a rent review or the lease ends. However, the gearing effect still applies to the market rent, and a traditional profit rent based approach will not reflect this in the valuation figure.

In conclusion there are significant problems with the traditional approach to leasehold valuation. While it is common to debate the validity of dual rate and dual rate tax adjusted approaches, the highly geared growth potential of the profit rent is a major issue which means that any valuation based on finding the profit rent is questionable. It remains to be seen how long it will take the profession to adopt an alternative approach to market valuation. In the meantime, the more modern methods can be used to find the worth of leasehold investments.

Further reading

Property Investment Appraisal, Baum and Crosby, 3rd ed, Blackwell, 2007

18 Principles of Property Investment

Valuers now recognise that there is a distinction to be made between a valuation as an objective exercise to estimate the price at which an asset will change hands in the open market, and a calculation of worth, which is a subjective view of the price an individual investor may be willing to pay.

This is an important distinction. If the market value of an asset exceeds the worth to the owner then it would make sense for that owner to realise the additional value by selling the asset. If, on the other hand, the market value of the asset is less than the owner's perception of its worth then the owner will retain it.

Baum and Crosby in their book *Property Investment Appraisal*, first published in 1988, argue that DCF analysis is particularly fruitful in arriving at estimations of worth precisely because this is a subjective and not a market based exercise. Investors are not necessarily investors in property per se; they will consider property as an option alongside a range of different investment opportunities. This requires a degree of cross investment comparison. This would be impossible using the market based all risks yield (ARY) which is uniquely applied to property and which is largely meaningless in any other investment context. Even in a property context they argue that if anything the all risks yield should be seen as a unit of comparison:

> "An explicit cash flow is merely the unbundling of factors that are hidden in the ARY approach. Where these factors are bundled together in the ARY, they cannot be identified separately and the yield therefore reverts to a unit of comparison rather than an investment indicator. Since this has no

> rational investment basis it can be identified only through market evidence of similar properties" Crosby N, *Estates Gazette* 14 October 2006, p216

A subjective estimate of worth may well start with a target yield or rate of return and this in turn may be linked to yields on alternative investments. DCF analysis can then be used to assess how the investment will perform compared to this target. If the target is met or beaten then it may make sense to acquire the asset, if the return is less than the target yield the investor should not acquire it, and if it is already held in the investor's portfolio disposal should be considered.

Using DCF as an investment analysis tool

We have already seen in Chapter 16 that one of the attractions of using full DCF is that it allows for a more explicit and more complex modelling of cash flows over time, so that it is possible to build in a number of variables. One of the limitations of conventional approaches to valuation is that it is based on a simplification of future cash flows. These additional variables may include costs as well as trends such as rental growth and depreciation and any of these variables can be made explicit to reflect the requirements and assumptions of the individual investor. One important assumption to be made is that the investor will hold the investment for a finite time, the holding period, after which it is assumed that the investment is sold at its then market value. Research has shown that a holding period of ten years is sufficient in most cases to produce a reliable appraisal, although the calculation should be performed over a long enough period to include a reversion to market rent unless this is a very long time away. This is of course a notional disposal and is a necessary assumption which prevents the cash flow analysis proceeding to infinity, although the latter can be done by capitalising the last tranche of income in perpetuity at an appropriate yield.

Baum and Crosby discuss the construction of target yields based on the comparatively risk free redemption yield on government stocks. In 1988 they concluded that such a process was dangerous and unreliable but it is interesting to note that now many investment analysts have the confidence to do precisely this in making investment decisions.

Distinctions between property and other investments

For many years property has been regarded as unique as an investment. This has led to the unquestioned assumption that the methods used for valuing property should be distinct and should be carried out by people qualified and experienced in the field of property valuation. These assumptions are being challenged as a result of pressure from investors who want to be able to make cross investment comparisons and who require more comprehensive information about the quality and performance of their investments. This has led to a degree of convergence in investment appraisal methodology and the more commonplace acceptance of discounted cash flow as a means of making informed investment decisions about a wide range of income generating investments including property. Part of this acceptance is the result of the widespread availability of technology and specialist and general software applications that facilitate rapid and flexible calculations. We should be aware there are still valuers practising today who can remember a time when electronic calculators, never mind desk top computers, did not exist. In 1988, for example, few if any valuers would have access to a computer of any description and even fewer would have been able to use a spreadsheet.

Accepting that there is a need for cross investment comparison within investment appraisal does not alter the fact that there are important distinctions to be made if we are to fully understand the true nature of property as an investment. There is widespread agreement among authors of property valuation and property investment texts about the differences between property and other investments as illustrated in the table overleaf.

Characteristics of property investments

Readers are referred to these authors for a detailed analysis but the key differences are summarised below.

Indivisibility and high unit value

Most individual properties have a high value. The average price of houses nationally was in excess of £220,000 in 2007 and many commercial properties will be valued in millions rather than hundreds

Hoesli & Macgregor	Baum & Crosby	Millington	Enever & Isaac	Scarrett
heterogeneity			heterogeneity	heterogeneity
fixed location				
unit value		indivisibility	indivisibility	indivisibility
finance				
long term			durability	
management	operating expenses		special problems of management	costs of management
inelasticity of supply			inelasticity of supply	demand
depreciation				
government intervention			government intervention	government intervention
psychic income	psychic income			
price determination				
illiquidity	liquidity, marketability and transfer costs		high cost of transfer	
	income and capital growth	income and capital security/ inflation hedge		growth and inflation
	tax efficiency			
	risk		special risks	
		high cost of transfer		ease of dealing
			imperfect knowledge	adequate information
			decentralised market	wide market

or thousands of pounds. This combined with the fact that property is largely indivisible (incapable of being divided up into smaller units) means that *direct* investment in property requires high levels of personal wealth or finance.

Management costs

A second commonly agreed distinctive feature of property investments is the relatively high cost of the specialised management required. This is concerned in part with the maintenance of the fabric of the building but also reflects the need to manage the legal relationships between owner and occupiers, requiring the negotiation of leases, rent reviews and lease renewals. As a consequence a much larger proportion of gross investment income will be swallowed up in management costs than is the case with most other investments.

Heterogeneity

This is a further widely recognised difference. Every property is unique, with its own distinctive features. Even where properties are very similar they are bound to occupy different sites. Most other investment types are more or less homogenous; stocks, shares and commodities for example.

Inelasticity of supply

As we saw in Chapter 5, the supply of property is fixed in the short term. This is a function of the lag in the development cycle and the length of time it takes to respond to changes in demand. This tends to result in large fluctuations in price.

Liquidity, marketability and costs of transfer

Property lacks liquidity compared to other investments. In other words because of the complexity of, the time taken and the high cost of the transfer process it is relatively difficult, time consuming and expensive to convert a property asset into cash.

Government intervention

Property is subject to extensive government intervention. Its very permanence makes it attractive as a tax base and there are also widespread restrictions as to the use of property especially through town planning legislation and zoning control. Sometimes this intervention can be quite unexpected and even ill considered. Recent examples include punitive changes in the rules concerning the rating of empty properties, changes in the Inheritance Tax thresholds and temporary changes in stamp duty.

Market differences

There a number of key differences in the way in which the property markets operate. Some of these flow from the other characteristics such as heterogeneity. Markets tend to be decentralised and made up of large numbers of buyers and sellers with imperfect or limited information. This also means that mechanisms for price determination are not as efficient as is the case with other investments. Understanding these markets tends to require specialist knowledge.

The characteristics of property outlined above represent a significant collective downside risk, so much so that you could be forgiven for wondering why anyone should choose to invest in property at all. The main reason of course is growth.

Growth

The upside of property investment is the tendency of most properties over the longer term to exhibit positive growth characteristics. Over time rents will tend to increase as the demand for different types of property increases and where rents increase there will also be a tendency for capital values to increase. This means that property can be regarded as a good hedge against inflation.

Psychic income

This final distinction noted by Baum and Crosby recognises "the appeal of property unmatched by the alternatives", the prestige value of holding property.

The whole point of looking at property investments in the wider context of investment is to facilitate a more rational approach to decision making which considers property as one of a number of possible investment opportunities.

Making investment decisions

Using this type of subjective, appraisal based analysis can be illustrated by reference to a simple example. Assume you are acting for an investor client, a pension fund for example, and you have been offered a prime freehold retail investment opportunity in the centre of a large northern city. It is currently on the market and you have been given the following details:

> The shop is let to a national multiple tenant at a rent of £125,000pa on full repairing and insuring terms on a new 10 year lease with a review/break after five years. The asking price is £2,775,000.

If you agreed to purchase the investment at the asking price your client would receive an initial return on capital of 4.5%. This is found by simply dividing the current income by the purchase price.

Income/Purchase price
£125,000pa/£2,775,000 = 0.04505 or just over 4.5%.

At first sight this does not appear to be a particularly attractive proposition, especially when all the risks of holding a property are taken into account. We have already seen for example that property as an investment is not especially liquid. It is neither easy nor cheap to buy and sell. It has relatively high operating expenses. It is prone to a wide range of risks such as tenant default, change in market prices over time and there may be structural risks affecting the physical integrity of the building, sector risks affecting all retail investments and locational risks such as the possibility of the development of a competing shopping centre nearby which may cause a shift in shopping patterns. However one of the principal benefits of property investment is the general tendency for rental values to increase over time (growth). This single positive tendency can be so significant a factor that it tends to outweigh all the disadvantages of investing in property.

If we assume, for example, that rents will grow at a modest average of 3% pa over the next five years, by the time the first review is reached the rent will have increased as follows:

Current MR	£125,000pa
× A in 5 years at 3%	1.1593
Rent at first review	£144,913pa

The return on capital will then be:
£144,913pa/£2,775,000 = 0.0522 (just over 5.2%).

Not only that but, as the rental value increases, the capital value will tend to increase as well and so the investor will enjoy capital growth as well as income growth.

It is probable that the original asking price of the investment (£2.775m) was fixed by reference to the market based all risks yield for prime retail investments (4.5%) and this would have been found by analysing similar transactions.

This is all very well. It assumes there is ample evidence to support this market analysis. It also assumes that our investor is only interested in property investment. However, it is often the case that there is little or no comparable market evidence to support the selection of an all risks yield. Furthermore our pension fund investor may want to compare the performance of this individual property investment with a range of different investment opportunities such as bonds, cash and equities.

Now consider a slightly different scenario. This time our pension fund investor client has £2.75 million to invest and is faced with two mutually exclusive options:

1. the prime retail property offered at an initial yield of 4.5% as set out above or
2. a block of medium dated government stock with a redemption yield of 4.5%.

Your investor client needs to be able to make cross investment comparison to inform the decision making process. This is more difficult because the yields for each of the two different investments actually measure different things.

The property yield (4.5%) is an initial yield and is based on market evidence. As we have seen it is the initial income divided by the capital value or market price of the investment. This may be based on market evidence but the initial yield will reflect all the risks of the investment, including the positive tendency for property incomes to enjoy future rental growth.

By comparison the government stock investment has none of the downside risk tendencies of property; it is liquid, cheap to manage and virtually risk free being backed by the government. On the other hand the 4.5% redemption yield for the government stock represents a fixed interest yield where there is no growth potential.

The two yields then, while numerically very close, are measuring very different investments and so cross investment comparison is impossible based on this information alone.

Going back to basics it is established that the yield of an investment is in effect the reward to the investor for the following:

- time preference
- inflation
- risk.

In 1988, Baum and Crosby argued (at least from a theoretical perspective) that if we could identify a risk free rate (RFR) this could represent the starting point for the construction of a target yield for property investment. This target yield would be the true yield, or internal rate of return, required by the investor to make any investment attractive when compared with a risk free investment such as government stock. They suggested that the target yield could be found by starting with the risk free yield and adding a margin to reflect the particular additional risk characteristics of property:

Target Yield = Risk Free Yield (RF) + Risk Premium (RP)

Here the risk free yield is taken as the redemption yield on conventional gilts. The risk premium can be broken down into the generic risks of investing in property (lack of liquidity and management costs) as well as the specific risks attached to a particular property. These could include tenant risk, property risk, sector risk and location risk.

Returning to our investment example the target yield for the shop might be constructed as follows:

Risk Free Yield based on Government Stock			4.5%
add for:			
Generic property risk	say	2.5%	
Specific property risk			
Location	say	1.0%	
Tenant	say	0.5%	
Building	say	0.5%	
			4.5%
Target Yield			9.0%

It is now possible to carry out a DCF investment appraisal to determine whether or not to acquire the shop at the current asking price.

Year	Income	Outlay/Resale	Cash Flow
0		–£2,775,000	–£2,775,000
1	£125,000		£125,000
2	£125,000		£125,000
3	£125,000		£125,000
4	£125,000		£125,000
5	£125,000		£125,000
6	£144,913		£144,913
7	£144,913		£144,913
8	£144,913		£144,913
9	£144,913		£144,913
10	£144,913	£3,733,063	£3,877,976
IRR			7.27%

The above analysis assumes that the initial outlay or purchase price is made at the beginning of the holding period and rents are assumed to be paid annually in arrears.

Resale takes place at the end of the 10 year holding period and the resale price is found by capitalising the rent (including estimated growth over the 10 year period) at the all risks yield in perpetuity:

Market rent	£125,000pa	
× A for 10 years at 3%	1.3439	
Estimated rent at end of holding period	£167,988	
× YP in perp at 4.5%	22.2222	
Exit Value		£3,733,063

Some surveyors would reduce the exit value to reflect the costs of a notional disposal.

This analysis shows that if rental growth is assumed at 3% pa the internal rate of return falls some way short of the target rate of 9%. In fact to meet this target rate, growth in the rent would have to be more than double. In these circumstances the advice to the client may well be to seek to reduce the asking price by negotiation and failing that spend the available £2.775 million on the government stock.

Further consideration of the use of yield construction can be found in the *Sportelli* case (*Earl Cadogan* v *Sportelli* [2007] RVR 314) which, among other things, concerned the deferment rate to be used in assessing the value of the final reversion in Leasehold Reform Act enfranchisement cases (where the tenants of houses and flats have the right to enfranchise, or buy the freehold or their properties). Evidence from financial experts was taken in assessing the required return on the investment (R) which was based on a Risk Free Rate (RFR) plus a market based premium (RP) minus asset growth (g).

In arriving at the RFR it was held that the nearest empirical equivalent was the long term average return on index linked gilts, taken at 2.25%. Real growth was taken to be an average rate over the long term of 2% and the property risk premium was assumed to be 4.5%. Thus the deferment rate for houses was taken to be:

$$\text{RFR} - g + \text{MRP}$$

$$2.25\% - 2\% + 4.5\% = 4.75\%$$

With an additional 0.25% for flats to reflect additional service charge, repair and management problems.

Theoretically it might be necessary to adjust for specific factors such as the length of the term, location and obsolescence although in this case these were considered to be already reflected in the vacant possession value of the reversion. (For a further discussion of yield construction in the *Sportelli* case see Crosbie (*sic*) & Pottinger, Estates Gazette 14 October 2006 pp216–218.)

Baum and Crosby concluded in 1988 "that the process of appraisal by ... yield construction is dangerous and impossible to practise". However there is evidence to suggest that this type of subjective approach to appraisal is now commonly used as a decision making tool by investment appraisers. Despite the subjectivity of the various yield adjustments shown above and the dangers inherent in trying to predict future growth, what can be said is that this type of analysis does make the investor and appraiser's assumptions more explicit rather than burying them in some mysterious market based measure which makes cross investment comparison difficult. Furthermore the more explicit approach to analysis can assist the decision making process.

What to use and when?

Chapter 8 introduced the investment method of valuation and in Chapters 14 through to 18 we have explored a range of different methods of valuing, and estimating the worth of, property investment opportunities. These are either conventional, growth implicit approaches based on the application of the all risks yield or growth explicit methods based on discounted cash flow and the internal rate of return. These growth explicit methods can themselves be divided into full DCF methods and short cut DCF methods which try to replicate the layout and general appearance of traditional growth implicit investment valuations. Even relatively experienced valuers faced with this rather bewildering array of possibilities when deciding how to capitalise a property based income flow can be excused some confusion.

Figure 18.1 Summary of investment valuation methods

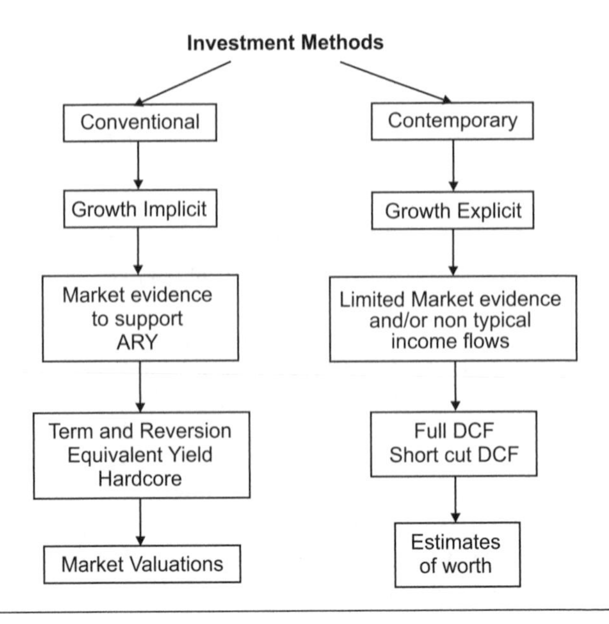

At one level the decision over which approach to choose is a matter of personal preference but perhaps it is better to approach the decision with the following guidance in mind.

- In cases where there is sufficient and reliable market evidence available on which to base an assessment of the all risks yield, conventional growth implicit approaches should be adequate. This should be the case with fully let freeholds on prime properties.
- The conventional hardcore method can be useful in the valuation of over rented properties where the top slice of income is significantly at risk.
- Where evidence is in short supply and significant intuitive adjustment is required, valuers should consider using growth explicit methods.
- For market valuations use growth implicit methods but for estimates of worth DCF is probably more appropriate because it provides the opportunity to assess the impact of different scenarios and assumptions.

Further reading

An Introduction to Property Valuation, Millington A, 5th ed, Estates Gazette, 2000

Property Investment Appraisal, Baum A and Crosby N, Routledge, 1988

Property Investment: Principles and Practice of Portfolio Management, Hoesli M and Macgregor B, Longman, 2000

Property Valuation: the five methods, Scarrett D, Routledge 2008.

The Valuation of Property Investments, Enever N and Isaac D, Estates Gazette, 2002

Index